Patrick Moore's Practical Astronomy Series

D1319634

Other Titles in this Series

(*continued after index*)

Aurora

Observing and Recording Nature's Spectacular Light Show

Neil Bone

 Springer

Neil Bone
'The Harepath'
Mile End Lane
Apulduram
Chichester
West Sussex PO20 7DZ
UK
bafb4@central.susx.ac.uk

Library of Congress Control Number: 2006940469

Patrick Moore's Practical Astronomy Series ISSN 1617-7185
ISBN-10: 0-387-36052-2   e-ISBN-10: 0-387-68469-7
ISBN-13: 978-0-387-36052-2   e-ISBN-13: 978-0-387-68469-7

Printed on acid-free paper.

springer.com

Contents

Preface

The popular astronomical literature has, over the years, been rather unkind to phenomena of astronomical origins occurring in the Earth's atmosphere. Together with meteors, the aurora has often been presented as something rather exotic, and not really astronomical anyway, before the author passes on to the next topic. It is my hope here to redress the balance somewhat, by presenting an account of the aurora, its causes, and how to observe it, in a form accessible to the reasonably well-informed amateur astronomer.

Many years ago, back in the late 1960s, I was as a young newcomer to astronomy and avid reader of the even-then numerous general introductions to the subject. Among these, one of the most eye-catching features would often be a garishly colored "artist's impression" of the Aurora Borealis or northern lights. These displays, one would usually be informed, could only be seen from high Arctic latitudes, and were caused by a rain of particles from the Van Allen belts. Our understanding of auroral phenomena has certainly moved on from then, and "Space Weather" has become a hot current topic, given its influence on satellite operations on which life in the early twenty-first century has become ever-more reliant. We also now have excellent photographic equipment capable of recording auroral displays in all their beauty and magnificence! Countless amateur astronomers and others marveled at, and photographed—and even digitally imaged—the "Hallowe'en Storms" of 2003, perhaps the most intense and extensive auroral activity of modern times.

When I saw my first aurora back in 1973, I mistook it for an unusually early moonrise! Simply, the pictures in the popular texts I had immersed myself in were inadequate preparation for the real thing. Here, I have assembled a set of photographs that should give ample guidance as to how the aurora really appears, and with familiar constellations often recorded in the background, the reader should also gain some appreciation of the scale of displays.

I was fortunate enough to be born and raised in Scotland, from where I enjoyed numerous opportunities to observe the aurora (not strictly a polar phenomenon at all!) subsequent to my February 1973 misidentification. Having moved, through professional commitments, to more southerly climes in England, I find the aurora a less-frequent visitor to my skies, but even from the depths of Sussex, I have over the past couple of decades witnessed several good displays. Events visible to southern England spark a lot of interest, and I have given countless talks on the subject to local astronomical societies, particularly in times following major displays. This book brings together much of the lecture material as a theoretical background to how the aurora comes about, as well as outlining how to go about recording such events for oneself.

Many people have helped me in compiling this account of the aurora. I should especially acknowledge the long-standing contributions of Ron Livesey—for 25 years

Director of the BAA Aurora Section—and Dr. Dave Gavine. Dave, and colleagues such as Richard Pearce, Tom McEwan, and Russell Cockman have generously allowed me to use copies of their frequently stunning images to illustrate my lectures over the years, and some of their work can be seen in this volume. The late Dr. Michael Gadsden inspired many to take up the related study of noctilucent clouds, covered in the final chapter.

I should also give a special mention to my wife, Gina, to whom this book is dedicated, for her patience and understanding during the time taken compiling this work. We have been lucky enough to observe together one or two of the more spectacular aurorae of cycles 22 and 23.

As I write, cycle 23 is drawing to its close, and auroral activity at lower latitudes has died away for the time being. In another couple of years, however, we may well expect the onset of some vigorous activity as the sunspots start to appear again in large numbers; some forecasts suggest we may even be in for a particularly good solar cycle for major aurorae, peaking around 2010. Time will tell, but for the moment, it is my hope that its readers will find this book a useful introduction to the aurora and how to observe it.

Neil Bone
Chichester, 24th September 2006

CHAPTER ONE

Atmospheric Phenomena

A clear, early spring evening, Thursday, 6–7 April 2000 had been keenly awaited for some months by amateur astronomers across northwestern Europe and North America. On this occasion, the planets Mars, Saturn, and Jupiter would form a neat grouping in the western sky together with the waxing crescent Moon in the growing twilight. Such conjunctions, when members of the Solar System are seen relatively close together in line of sight, are not especially rare, but the apparent gathering of four objects in a small area of the western evening sky was sufficiently noteworthy and photogenic to attract a lot of attention and, visible for that one evening only, was an event to be savoured and imaged while it lasted! Clear skies across the British Isles afforded excellent viewing and were especially welcome after a spell of cloudy, unseasonably cold weather.

As the twilight deepened and the planetary–lunar grouping sank low in the northwest, observers began to notice something odd about the sky. Patches of red light appeared across the north and northwest, resolving at times into curtained and banded forms: as an unexpected bonus to an already splendid evening, the aurora had returned to low-latitude skies after some years of absence.

Triggered by violent magnetic activity on the Sun a couple of days previously, the geomagnetic storm and its associated extensive auroral display continued throughout the night, and was also enjoyed by American observers. The first of several such events in sunspot cycle 23, the major aurora of 6–7 April 2000 was visible down to the southern United States, and from locations such as Portugal in the south of Europe where the aurora is seldom seen. To the thousands who witnessed it, this spectacular natural light show certainly made this a night to remember.

Part of the thrill lay in the unexpectedness of the event; while the conjunction was long-anticipated (desktop planetarium programs can predict such alignments for centuries in the future with consummate ease), the aurora took many—even

Figure 1.1. The strongly coloured aurora of 6–7 April 2000 was widely seen across the British Isles. The Author captured this view from his home near Chichester, West Sussex.

professional scientists—by surprise. Recent developments afford the possibility of more reliably forecasting the likely occurrence of aurorae, and an important part of what follows in this book will discuss the online resources where the reader can obtain information in real time about conditions in near-Earth space and whether auroral storm conditions are likely.

Aurorae in History

Aurorae have been observed and recorded throughout history, being a source of awe to many and an inspiration for myth and poetry. There is some debate as to where the descriptive name of *aurora borealis*—in Latin, "northern dawn"—originated. Some writers ascribe the first use of the term to the French astronomer Pierre Gassendi (1592–1655) following a display seen on 12 September 1621. Others give the priority to Galileo (1564–1642)—famed, of course, as the first to apply the telescope to astronomical purposes—who also witnessed the 1621 aurora. Perhaps more realistically, aurora borealis has its origins with Gregory of Tours more than a millennium before either Galileo or Gassendi. Gregory (538–594) was involved in clerical and political life, and wrote the ten-volume *The History of the Franks* (French). In this are clear descriptions of a number of celestial phenomena such as parhelia. Several passages can be interpreted as accounts of auroral displays:

"While we were still hanging about in Paris portents appeared in the sky. Twenty rays of light appeared in the north, starting in the east, and then moving round to the west. One of them was longer than the others and shone high above them: it reached right up into the sky and then disappeared, and the others faded away, too." (AD 578)

"While I was staying in Carnignan, I twice during the night saw portents in the sky. These were rays of light towards the north, shining so brightly that I had never seen anything like them before: the clouds were blood-red on both sides, to the east and to the west. On a third night these rays appeared again, at about seven or eight o'clock. As I gazed in wonder at them, others like them began to shine from all four quarters of the earth, so that as I watched they filled the entire sky. A cloud gleamed bright in the middle of the heavens, and these rays were all focused on it, as if it were a pavilion the coloured stripes of which were broad at the bottom but became narrower as they rose, meeting in a hood at the top. In between the rays of light there were other clouds flashing vividly as if they were being struck by lightning. This extraordinary phenomenon filled me with foreboding, for it was clear that some disaster was about to be sent from heaven." (AD 586)

The "pavilion" described above sounds remarkably like the coronal form adopted by the aurora in the most major storms when activity extends overhead and on toward the equatorwards side of the sky.

Among Gregory's descriptions, one particularly stands out as a likely original for the nomenclature:

"At this time there appeared at midnight in the northern sky a multitude of rays which shone with extreme brilliance. They came together and then separated again, vanishing in all directions. The sky towards the north was so bright that you might have thought that day was about to dawn."

Aurorae in Mythology

The aurora has, not surprisingly, often entered into the folklore of peoples living at higher latitudes. The Scottish city of Aberdeen is connected to the Northern Lights or "Heavenly Dancers" in a popular song, for example. Robert Burns, Scotland's national bard, makes mention of the aurora in his galloping epic Tam O'Shanter:

> But pleasures are like poppies spread,
> You seize the flow'r, its bloom is shed,
> Or like the snow falls in the river,
> A moment white—then melts for ever,
> Or like the borealis race,
> That flit ere you can point their place;
> Or like the rainbow's lovely form
> Evanishing amid the storm-
> Nae man can tether time nor tide;
> The hour approaches Tam maun ride
> *The Edinburgh Edition, 1793*

The much-traveled poet, Robert William Service (1874–1958) spent some time in the far north of Canada, during the declining years of the late nineteenth century Klondike and Yukon gold rush. Some of his writings provide a fascinating record of frontier life. Not surprisingly, the aurora occasionally features as a backdrop:

> There where the mighty mountains bare their
>   fangs unto the moon;
> There where the sullen sun-dogs glare in the
>   snow-bright, bitter noon,
> And the glacier-gutted streams sweep down at
>   the clarion call of June:
>
> There where the livid tundras keep their tryst
>   with the tranquil snows;
> There where the silences are spawned, and the
>   light of hell-fire flows
> Into the bowl of the midnight sky, violet, amber
>   and rose
>
> There where the rapids churn and roll, and the
>   ice flows following run;
> Where the tortured, twisted rivers of blood rush
>   to the setting sun—
> I've packed my kit and I'm going, boys, ere
>   another day is run.
>   *The Heart of the Sourdough*

A longer narrative tells the tale of three miners seeking their fortune in the far north, guided by their dreams. The sole survivor—a down and out—relates their experiences and describes the aurora:

> Oh, it was wild and weird and wan, and ever in
>   camp o' nights
> We would watch and watch the silver dance of
>   the mystic Northern Lights.
> And soft they danced from the Polar sky and
>   swept in primrose haze;
> And swift they pranced with their silver feet,
>   and pierced with a blinding blaze.
> They danced a cotillion in the sky; they were
>   rose and silver shod;
> It was not good for the eyes of man—'twas a
>   sight for the eyes of God.
> It made us mad and strange and sad, and the
>   gold whereof we dreamed
> Was all forgot, and our only thought was of the
>   lights that gleamed.
>
> And the skies of night were alive with light, with a
> throbbing thrilling flame;

> Amber and rose and violet, opal and gold it came.
> It swept the sky like a giant scythe, it quivered
>   back to a wedge;
> Argently bright, it cleft the night with a wavy
>   golden edge.
> Pennants of silver waved and streamed, lazy
>   banners unfurled;
> Sudden splendors of sabres gleamed, lightning
>   javelins were hurled.
> There in our awe we crouched and saw with our
>   wild, uplifted eyes
> Charge and retire the hosts of fire in the
>   battlefield of the skies.

The storyteller informs us that the aurora originates from a hollow mountain range on the polar rim. Echoing a popular belief among the frontier gold miners of the time, he finally reveals:

> Some say that the Northern Lights are the glare
>   of the Arctic ice and snow;
> And some say that it's electricity, and nobody
>   seems to know.
> But I'll tell you now—and if I lie, may my lips
>   be stricken dumb—
> It's a mine, a mine of the precious stuff that men
>   call radium.
>
> *The Ballad of the Northern Lights*

References to auroral displays can be found in ancient literature. Perhaps the earliest record, unearthed by Durham astronomical historian Prof. Richard Stephenson, dates back to Babylonian tablets from March 567 BC. An oblique reference to the aurora, also dated to the sixth century BC, is made in the Bible:

> "And I looked, and, behold, a whirlwind came out of the north, a great cloud, and a fire infolding itself, and a brightness was about it, and out of the midst thereof as the colour of amber, out of the midst of the fire."
>
> Ezekiel 1:4

Over the years there has been much speculation as to the possible astronomical nature of the Star of Bethlehem. Planetary conjunctions, novae, and comets have all been proposed as the celestial events interpreted by the Biblical Magi as a sign that a new King of Israel had been born. It is quite possible that the aurora might provide a further, reasonable, alternative. Computations extrapolating the ephemerides back to the time of Christ indicate that there were no bright planetary conjunctions at the appropriate time, while searches through contemporary Oriental records give no indication of a candidate comet or nova. The rare penetration of auroral activity to the latitudes of the Middle East—a once-in-a-lifetime event, perhaps—would certainly be sufficiently unusual to be noted by such watchers of the sky as the Magi. In its coronal form, the aurora may very well assume the appearance of a stylized "star," with rays and other forms radiating out from a central point.

Norse mythology makes frequent reference to the bridge Bifrost, a burning, trembling arch across the sky, over which the gods could travel from Heaven (Asgard) to Earth. It is not unlikely that the inspiration for the bridge was the aurora. In a parallel to Bifrost, Finnish mythology refers to a river—Rutja—which stood in fire, and marked the boundary between the realms of the living and the dead.

The vivid red sometimes seen in intense auroral displays can probably be associated with the Viking "vigrod," or war-reddening.

In Scandinavian mythology, the Valkyries, "Choosers of the Slain," were beautiful young women mounted upon winged horses. Their role was to visit battlefields to select the bravest of those who had fallen (the Einherjar), and escort them to Odin's Hall (Valhalla), in preparation for the impending battle of Ragnarok. In some traditions, auroral rays were perceived as lights reflected from the Valkyries' armor as they rode the sky.

Some Norwegian folklore describes the aurora as a harbinger of harsh weather: snow and wind are believed to follow bright displays. Another Norwegian folk-legend suggests that the aurora is a celestial dance by the souls of dead maidens. A Swedish tradition associates the aurora with light reflected from the scales of fish in large shoals—Sillblixt, or "herring flash."

Other Scandinavian folk tales talk of a contest among the swans to see which could fly farthest north. Some competitors became trapped in the northern ice, and as these swans flap their wings in a bid to escape, the light of the aurora results.

The Norse chronicle *Kongersepeilet* (The King's Mirror) from about 1230 AD makes what is probably the first explicit reference to the Northern Lights (Nordrljos). Here, they are described as rays of sunlight reaching over Greenland, then thought to represent the edge of the flat, ocean-surrounded Earth:

> "[The northern lights] resemble a vast flame of fire viewed from a great distance. It also looks as if sharp points were shot from this flame up into the sky, these are of uneven height and inconstant motion, now one, now another darting highest; and the light appears to blaze like a living flame."

Eskimo peoples in the Hudson Bay area of North America, and elsewhere, are naturally very much aware of auroral phenomena. A common belief among the Eskimos is that the aurora can be attracted by whistling to it, while a handclap will cause it to recede. Other Eskimo beliefs suggest that the aurora is produced by spirits, playing a game of celestial football with the skull of a walrus. (One group, on Nanivak Island, suggested that a human skull was, instead, used by walrus spirits!). Some Eskimo groups regard the aurora as an indicator of good weather to be brought by the spirits. Alaskan Eskimos at Point Barrow saw the aurora as malevolent, and carried weapons for protection if venturing outside when it was present. It is also said by some Eskimos that:

> "He who looks long upon the aurora soon goes mad!"

Some tribes of North American Indians believed the aurora to be the light of lanterns carried by spirits seeking the souls of dead hunters. Like the Point Barrow Eskimos, Fox Indians in Wisconsin feared the aurora, seeing in it the ghosts of their dead enemies. Other tribes perceived the aurora as the light of fires used by powerful northern medicine men.

The aurora has also entered the folklore of the Australian aborigines, who saw it as the dance of gods across the sky. To the Maoris of New Zealand, the aurora is Tahu-Nui-A-Rangi, the great burning of the sky.

Aurorae may well have been the source of Chinese dragon legends. The twisting snake-like forms of active auroral bands are often portrayed as celestial "serpents" in ancient chronicles. European dragon legends, too, may have their origin in auroral activity, although some commentators also ascribe these to descriptions of meteoric fireballs.

In ancient Roman and Greek records, references may sometimes be found to "chasmata" in the sky, the auroral arc structure being regarded in such instances as being the mouth of a celestial cave. The term *isochasms* is used nowadays to relate two geographical points that share an identical frequency of auroral occurrence.

Even in quite modern times, misconceptions regarding the cause of the aurora have persisted among the general public. Writing in *National Geographic* in 1947, for example, American auroral scientist Charles W. Gartlein relates how, in his youth, people in mid-west America widely believed the aurora to be the reflection of sunlight from the polar ice, disregarding the perpetual darkness of the winter months at Arctic latitudes! Another romantic notion, long-since safely dismissed, was that auroral light results from icebergs crashing together in the polar seas.

Modern scientific understanding of the processes underlying the aurora is now sufficiently advanced that good working models to describe the causes of the polar lights are available. This understanding, however, does not detract from the majestic spectacle that a major auroral display can present and—rather as is the case with total solar eclipses, thunderstorms, and other displays of Nature at its grandest—an active, colorful display of the Northern (or Southern) Lights can still trigger, even in an informed observer, primitive emotions of wonder.

Other Atmospheric Phenomena

The aurora is the most awe-inspiring of a range of phenomena occurring in the atmosphere, many of which become familiar to regular watchers of the skies. Amateur astronomers who carry out (or at least attempt to carry out) regular observations are all too familiar with the various cloud forms in the lower atmosphere's "weather layer"—the troposphere. Clouds all too often disrupt a night's plans, or obscure that once-in-a-lifetime event like a total solar eclipse or meteor storm. In a maritime temperate climate like that of the British Isles, the weather is often quite dynamic, bringing frontal systems and their attendant cloud at intervals of a few days.

Haloes

While the arrival of high, thin cirrus clouds (at altitudes of around 10 kilometers) ahead of a frontal system can signal the loss of planned astronomical observations, the optical phenomena produced by refraction of sunlight in the ice crystals of which this cloud is comprised can be of interest in their own right. The tiny ice crystals in

Figure 1.2. Among the commonest optical phenomena seen when ice-crystal cirrus clouds are present are parhelia, or Sun-dogs, like this one imaged low in the evening sky in August 2005 by the Author.

cirrus have a hexagonal geometry, and under the right conditions each can act as a minute prism through which sunlight (or moonlight) can be refracted. A consequence of this is the occurrence of the family of halo phenomena, which can sometimes be seen—usually when the Sun is low in the evening or morning sky—on occasions when the cirrus sheet is extensive.

Most commonly seen are the often colorful brightenings known as *parhelia*—mock Suns or Sun-dogs. The laws of refraction dictate that these appear at an angular distance on the sky of 23 degrees from the Sun, and at more-or-less the same altitude (elevation) above the horizon as the Sun. Sometimes there may be only a single parhelion, either to the Sun's east or west. On other occasions, both are present. Appearing like little patches of outside-in rainbow (with the red closest to the Sun, blue farthest), parhelia are usually brightest when the Sun is low in the sky; they are probably most often noticed in the evening sky as the Sun is setting, and would-be observers make a check to see whether conditions might be clear later on.

Given the right conditions, with an extensive, fairly even cirrus sheet covering much of the sky, the parhelia can be seen as brightenings just outside the circumference of a *halo* surrounding the Sun at a distance of $23°$ (radius $23°$, diameter $46°$). Directly above the Sun, another brightening—the *upper tangent arc*—may also be apparent on the halo's circumference.

Other components of the halo family of refraction phenomena include a *parhelic arc*, extending through the parhelia, parallel to the horizon at the same altitude as the

Figure 1.3. Under favourable conditions, the complete 23° halo may be seen around the Sun when cirrus cloud is present. Image: Neil Bone.

Sun. When present, this is usually seen only as a partial streak at temperate-latitude locations; from the Antarctic, where cirrus fields can be much more extensive and stable, the parhelic arc's complete 360° extent has been seen and photographed.

The halo family does not stop here! A further pattern of light refraction within ice crystals in the high atmosphere results in a second, outer set of halo phenomena at an angular distance of 46° from the Sun under the right conditions. The 46° radius halo is color-reversed relative to that at 23°, with blue on the sunward side, red on the outside. A bright, often strongly colored *circumzenithal arc*—tangential to the upper part of the 46° halo—can sometimes be seen, and parhelic brightenings may also be evident, level in altitude with the Sun and just outside the 46° halo. The circumzenithal arc is prominent only when the Sun is low, and its appearance requires a solar elevation of less than 32° above the horizon; at midsummer when solar elevations in excess of 58° may be attained, a *circumhorizontal arc* can sometimes be seen 46° below the Sun.

All these halo phenomena can be imaged, sometimes making spectacular, colorful pictures. With the camera being aimed toward the dazzling Sun, care has, of course, to be taken. As with any other piece of optical equipment, the observer should avoid direct viewing of the Sun through the camera viewfinder. One way to ensure this—and enhance the resulting image—is to hide the Sun behind a suitable local obstruction—a tree, the corner of a building or whatever; amateur astronomers afflicted by light pollution will enjoy the irony of using streetlights for this purpose!

For imaging of halo phenomena in a brightly sunlit sky, slow film (ISO less than 100), or a slow setting on a digital camera, is preferred; there is usually more than

Figure 1.4. Circumzenithal arcs are often highly coloured (inexperienced observers sometimes describe them as 'partial rainbows'), and can be bright especially in the evening when the Sun is low. Image: Neil Bone.

enough light. A narrow lens aperture will be sufficient. If using the camera's internal meter to set the exposure, a good rule of thumb is to underexpose by one or two f-stops. This will darken the background sky somewhat, better showing the halo or Sun-dog.

Closer examination of the resulting image can be interesting, revealing, for example, that the sky inside the 23° halo is markedly less bright than that outside—a result of preferential refraction of sunlight away from this region.

While most obvious when produced by a low Sun, equivalent halo phenomena can also be produced by the Moon in the night-time sky. However, as moonlight is much less intense, the *parselene* (mock Moons) and halo are fainter: only the 23° radius family members are likely to be seen, and then only when the Moon is at its brightest, within about five days of Full. Again, these can be attractive targets for imaging, if less colorful than their solar equivalents.

Sun-Pillars and Related Phenomena

Sometimes, on cold winter afternoons, the setting Sun will be seen to have extending upward from it perpendicular to the horizon a faint streak of light, perhaps slightly fanned-out toward its upper edge. Such Sun-pillars form in calm conditions, when ice crystals hang suspended in the lower atmosphere. The best-understood mechanism for formation of Sun-pillars involves refraction of sunlight through flat, hexagonal "plates" of ice, floating with their flat surface parallel to the horizon.

At night, under the same conditions, streetlights may also produce pillars, and these can occasionally be bright and colorful. Multiple vertical pillars from artificial light sources have, in the past, been mistaken for auroral displays. As we shall see below, however, the vertical ray structure characteristic of active, "curtain" aurora is seldom static for a prolonged period—unlike the stationary light pillars. Experienced observers will watch for lateral movement and changes in brightness as diagnostics of genuine aurorae.

Rainbows

In addition to the ice crystal phenomena of haloes and pillars, there are a number of optical phenomena associated with water droplets in the troposphere. The colored rings of the corona, produced around the bright Moon as a cloud passes in front of it, result from diffraction of light by uniformly small water droplets. Another diffraction phenomenon, seen around the edges of clouds close to the Sun, is iridescence, producing red and green fringes. Most familiar of all, however, must be the rainbow.

Rainbows are commonly seen in showery conditions, often at the end of the day as the Sun begins to break through the clouds. Produced by internal reflection of light through spherical raindrops back toward the observer, rainbows appear opposite the Sun in the sky. The laws of reflection dictate that emerging rays of light appear in preferred directions away from the anti-solar point in the observer's field of view. The brightest set appear at an angular radius of 40–42 degrees from the anti-solar point. Passing through the droplet, light of longer wavelength is refracted to a greater extent than short, resulting in the familiar gradation from red on the rainbow's outside, to blue on the inside.

This pattern applies to the bright, *primary* rainbow, produced by a single internal reflection.

Commonly, a *secondary* bow, resulting from double reflection of light within the raindrops, will be seen at a greater radius (around $51°$) from the anti-solar point. In the secondary bow, the order of colors is reversed, with red on the inside, blue outside. Secondary bows are fainter than primaries.

It hardly needs saying that rainbows make attractive photographic subjects! Examination of images, or even of rainbows in real time, will show that the sky inside the primary bow is brighter than that outside. Also, the sky is brighter outside the secondary. This is, again, a consequence of the preferential reflection of sunlight in certain directions, and away from others. The region between the primary and secondary bows is sometimes referred to as Alexander's Dark Band, named for the Greek Alexander of Aphrodisias, who first described it around AD 200.

Rainbows appear at their most extensive when the Sun is low, placing the anti-solar point higher in the observer's sky. For this reason, and since showers are most frequent in late afternoon, early evening rainbows—perhaps heralding clearance to good night-time skies!—are often the most impressive.

Figure 1.5. Primary (left) and secondary rainbows on a showery Edinburgh evening. Image: Neil Bone.

Meteors

From the astronomical perspective, clouds and their related optical phenomena in the atmospheric "weather layer"—the troposphere, extending to altitudes of 10–15 km above sea level—are, of course, mainly a distraction. One might look on even a very fine halo display as merely compensation for the loss of a potential night's observing.

The higher reaches of the Earth's atmosphere are, however, a realm where phenomena of more direct astronomical interest do occur. These include the aurora, which is this book's main topic, as well as other more exotic clouds to which we shall return in a later chapter. Another phenomenon of patently astronomical origin which occurs in the most tenuous parts of the upper atmosphere, at altitudes typically between 80 and 100 km is *meteors*.

While the atmosphere at these altitudes is a good approximation to a laboratory vacuum, with a density of only a millionth that at ground level, this is sufficient to destroy small particles of incoming interplanetary debris arriving at velocities between 11 and 76 km/sec. An estimated 16,000 tonnes of this material, most of it in the form of small, low-density dustballs, is swept up by Earth each year. Traveling at high velocities, the small particles—*meteoroids*—are destroyed by a process of ablation as they collide with atoms and molecules of the upper atmosphere. The friction results in the meteoroid giving up its kinetic energy as heat, ionization, and the short-lived streak of light seen in the night sky as a meteor. Most meteors have a visual duration of a couple of tenths of a second. The brighter examples—typically produced by larger meteoroids—may leave behind a briefly-lingering fading ionization train, which can last for a few seconds or, rarely, tens of seconds after the meteor itself has gone.

Many basic astronomy texts repeat the semi-myth that around six or seven meteors can be seen each hour by an attentive observer. In reality, there is considerable variation in meteor activity over the year. During the spring months, the background *sporadic* meteor activity may be as low as one or two per hour, although it reaches double figures in the autumn. When one of the major annual showers like the Perseids or Geminids is close to its peak, rates in excess of 60 per hour are sometimes obtained by observers at clear, dark locations.

Most of the annual showers are produced by trails of debris—meteor streams—laid down by comets. The meteoroids are shed from cometary nuclei while these are close to perihelion, and over several successive returns to the inner Solar System a comet may produce an extensive stream: those of comets 1P/Halley and 2P/Encke, for example, cover considerable swathes of the inner Solar System. The annual showers recur at the same time each year, as Earth returns to the position where its orbit intersects that of the meteor stream. During a shower, the attentive observer will notice that, in addition to the trickle of random sporadic background meteors, many meteors appear to emanate from a single part of the sky. This is the shower *radiant*, and the effect is one of perspective—shower meteors have parallel tracks in the upper atmosphere, but appear to diverge from a small area reflecting the intersection of the stream orbit and that of the Earth.

Observationally, the radiant effect is very useful, allowing meteors to be identified as coming from a given source. The main showers are named for the constellation in which their radiant lies—the Perseids come from Perseus, Geminids from Gemini and so forth.

Simple visual observations consist of making meteor watches, preferably in intervals of an hour, or multiples of an hour, at a time. The equipment requirements are minimal—the wide field of view of the naked eye is exploited by an observer, comfortably seated in a recliner and warmly wrapped against the elements. Recording is done by pen on paper by illumination of a red flashlight (to preserve night vision), or sometimes to a simple battery-powered voice recorder. For it to be of use in later systematic analysis, the watch's start and end times (in Universal Time, UT) should be recorded, and it is also important that the observer should give an indication of sky transparency (normally quoted as a naked eye stellar limiting magnitude).

During the watch, the observer records time of appearance (usually to the nearest minute), magnitude using background stars or the planets as comparisons, and type (shower association, or whether from the sporadic background) for each meteor. A meteor is assigned shower membership if its path can be projected *backwards* across the sky to the position of a known radiant expected to be active on the night. Other useful details include the presence and duration of any persistent ionization train, pronounced color, fragmentation in flight, and relative speed (slow/medium/fast).

Watch data, pooled from several observers, are analyzed to derive a Zenithal Hourly Rate (ZHR), corrected for sky transparency and the angular elevation of the shower radiant: all else being equal, when the radiant is low, the observed rates will be pegged

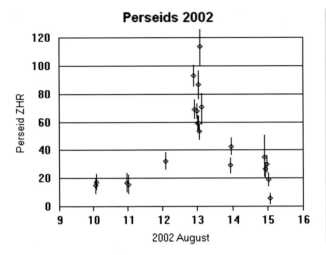

Figure 1.6. Activity profile for the Perseid meteor shower in 2002, based on BAA data. The usual sharp peak on 12–13 August is evident.

back somewhat. Figure 1-6 shows an analysis of British Astronomical Association (BAA) results from the 2002 Perseids, as an example. Activity in this shower rises steadily through early August, then a more rapid increase from 10 to 11 August leads to a sharp maximum—typically at ZHR around 80—on 12–13 August. The subsequent decline from peak activity is rapid, and the shower is over by 20 August.

Simple visual observations of this kind, repeated year on year whenever moonlight and weather conditions permit, allow detection of long-term trends in the activity of meteor showers In the early 1990s, for example, the Perseids showed an additional, short, sharp peak with ZHR much higher than the regular maximum—a result of Earth encountering a more concentrated "filament" of debris in the stream relatively close to the parent, Comet 109P/Swift-Tuttle, which came back to perihelion in its 135-year orbit in 1992. Among other showers, the Geminids—interestingly, associated with an asteroid (3200 Phaethon) rather than a comet—have shown a gradual increase in peak activity and a broadening of their highest-activity period in the past couple of decades, a trend well monitored by amateur visual observers.

Of course the most dynamic changes in meteor shower activity in recent times have come in the Leonids, associated with Comet 55P/Tempel-Tuttle. Historically, the Leonids have provided great storms of meteor activity at intervals close to 33 years, most notably in 1799, 1833, and 1966. The 1998 return of the parent comet was keenly anticipated, and was associated with a marked upturn in Leonid activity from its regular fairly modest annual pattern beginning in 1994. A night of unexpectedly high activity on 16–17 November 1998 was graced by numerous fireballs: by established definition, a fireball is a meteor brighter than the planet Venus, i.e., magnitude −5 or brighter. A storm with ZHR briefly reaching 3000 was seen on 17–18 November 1999. Modeling of enriched debris filaments in the Leonid meteor stream by David Asher (Armagh Observatory, Northern Ireland), and Robert McNaught (Siding Spring Observatory, Australia), successfully predicted further storm outbursts, two each on 18–19 November 2001 and 18–19 November 2002. The second outburst in 2001 was the most intense of the series, reaching ZHR 5000 for a time—very impressive for

Figure 1.7. Two bright Leonid meteors caught in a 5-minute undriven exposure on Kodak Elite 400 ISO film, using a 50 mm lens at f/2.8, 16–17 November 1998, 0411–0416 UT. Image: Neil Bone.

those fortunate to witness it, but still some considerable way short of the estimated 200,000 ZHR that produced awe and terror over the western hemisphere on 12–13 November 1833. From 2004, Leonid activity appears to have settled back toward its more sedate annual pattern.

Many observers attempt photography during the more active annual showers. Typically, only those meteors brighter than magnitude 0 to −1 are likely to be recorded, and then only if fast films and wide lens apertures are used. A good set-up is to use ISO 400 film (black and white or color), with a standard 50 mm or wide-angle 28 mm lens at f/2.8 or faster. The camera, firmly mounted on a tripod, is aimed around 20–30 degrees to one side of the radiant at an elevation of 50 degrees above the horizon, and time exposures of 10–15 minutes' duration taken. The camera can be driven to follow the stars as Earth rotates, but this is not absolutely essential, and many are quite happy with undriven images on which the stars appear as short trails. When recorded, meteors will appear as long streaks across the star background.

Positional data can be derived from such work, and if two or more observers at well-separated locations (a distance of at least 20 km is recommended) capture the same meteor photographically, it may even be possible to obtain meteor start and end heights by triangulation.

Table 1.1. Principal annual meteor showers

Shower	Activity Limits	Peak	Peak ZHR	Radiant RA	Radiant Dec
Quadrantids	Jan 1–6	Jan 3–4	100	15h 28m	+50°
Lyrids	Apr 19–25	Apr 21–22	10	18h 08m	+32°
η Aquarids	Apr24–May 20	May 4–5	40	22h 20m	−01°
δ Aquarids N	Jul 15–Aug 20	Aug 6–7	10	23h 04m	+02°
δ Aquarids S	Jul 15–Aug 20	Jul 28–29	20	22h 36m	−17°
α Capricornids	Jul 15–Aug 20	Aug 2–3	5	20h 36m	−10°
Perseids	Jul 23–Aug 20	Aug 12–13	80	03h 04m	+58°
Orionids	Oct 16–30	Oct 20–22	25	06h 24m	+15°
Taurids N	Oct 20–Nov 30	Nov3–4	5	03h 44m	+22°
Taurids S	Oct 20–Nov 30	Nov 3–4	5	03h 44m	+14°
Leonids	Nov 15–20	Nov 17–18	Var	10h 08m	+22°
Geminids	Dec 7–16	Dec 13–14	100	07h 32m	+33°
Ursids	Dec 17–25	Dec 22–23	10	14h 28m	+78°

Digital cameras are now being applied successfully to recording meteors. Exposures set to ISO 800 or even ISO 1600 can capture meteors, but many systems are limited to a shutter-open time of only 30 seconds. In an automated system, this leads to 120 exposures in an hour, with the attendant data storage and battery lifetime issues. In time, however it seems inevitable that digital cameras will become better suited for this sort of work, and as in other areas of astronomical imaging, conventional film will be supplanted.

Serendipity certainly plays a large part in successful meteor photography, a point brought home during the major auroral display of 8–9 November 1991. At that time of year, the Taurids (associated with Comet 2P/Encke) are active, a shower which in some years (2005 was noteworthy) can produce numerous bright meteors and fireballs. A good example came on the night of a major, extensive aurora on 8–9 November 1991. On this occasion, several observers in parts of western England—aiming, of course, to photograph the aurora—had their shutters open for a few seconds at just the right moment (22h 55m UT) to capture a slow, bright, flaring Taurid meteor in the northern sky!

Meteors occur in the lower parts of what we might consider the "auroral" region of the atmosphere. Although quite different in nature, the two phenomena attract a great deal of interest from amateur astronomers, and have in common the advantage that the naked eye or simple photography represents the ideal means of observation. Indeed, a large number of the reports of auroral activity received by the BAA come from observers out under the open sky carrying out meteor watches.

Space Weather

Our atmosphere is, of course, essential to life on Earth. Without its natural "greenhouse" warming effect, Earth would be a subfreezing ball of lifeless rock. Oxygen-liberating photosynthesis, evolved in blue-green algae around 2 billion years ago, modified the early terrestrial atmosphere, leading eventually to that which we find benign today.

We are all affected by weather phenomena in the lower atmosphere. Dearth or excess of rain, localized storms or tornadoes, or the more widespread devastation visited by hurricanes all play a part in human affairs. Do the more astronomically connected phenomena of the high atmosphere have any significant influence?

Prior to the Leonid meteor storms of 1999, 2001, and 2002, there was considerable speculation in the popular media that artificial satellites—so vital for communication and navigation among numerous other roles—would be "sandblasted" or knocked out of their operational orbits by swarms of small dust particles in the 55P/Tempel-Tuttle debris stream through which Earth was to plough. In the event, no reports of damage emerged!

Satellites are, however, at risk during the intervals of enhanced solar activity that trigger major geomagnetic disturbances, and with which extensive auroral displays are associated. Strong electrical currents flowing at orbital altitudes during these storms can damage satellite systems, as can the impact of high-energy accelerated subatomic particles. "Bit flips," where information stored in a satellite's onboard computer may become altered, are a problem. Unexpected—and unwanted—engine firings resulting from software glitches can take satellites out of position, and deplete limited fuel reserves required for future maneuvering.

Several satellites have been lost to adverse conditions during geomagnetic storms: casualties include the Marecs-B marine navigation satellite in February 1982, and the Canadian communications satellites Anik E1 and Anik E2, lost during a major disturbance in January 1994.

The effects of geomagnetic storms can propagate to ground level. Deviations of magnetic compasses from their normal behavior during major storms were, in past times, a hazard for marine navigation. Electrical currents produced at ground-level during major geomagnetic events are a significant problem, particularly for those living at higher latitudes. Ground Induced Currents (GICs) during the March 1989 storm, for example, produced surges in the Quebec power grid, leading to a 9-hour electrical blackout in the Canadian province.

X-ray and ultraviolet radiation from the sunspot-forming solar regions in which the flares and coronal mass ejections responsible for geomagnetic mayhem arise also have an important effect in heating the Earth's tenuous outer atmosphere. At times of high solar activity, the atmosphere expands to higher altitudes, resulting in increased drag on orbiting satellites. This is sufficient to cause loss of orbital momentum, and unless this is corrected for by a rocket burn, any affected satellite will gradually sink lower, eventually re-entering the atmosphere and being destroyed. Among the notable victims of this effect was NASA's Skylab, in 1979—ironically, given the role that the space station played in first bringing to light some of the forms of solar activity that contributed to its demise. Similarly, the Solar Maximum Mission satellite met its end as a result of increased atmospheric drag in 1990.

The loss of satellites and damage to electrical grid systems is, of course, costly. In recent decades, as human activity has become more dependent on satellite technology, increasing priority has been placed on understanding the causes of geomagnetic storms and, where possible, forecasting them. The field of Space Weather is important in this sphere: advance warning of an impending geomagnetic storm might allow sensitive satellites to be placed in "safe" mode, for instance. In the future, proposed manned missions to Mars, involving long-duration trips in the potentially hazardous inter-planetary environment, will also, surely, rely on accurate Space Weather forecasting. The violent events at the Sun's surface that trigger geomagnetic storms are accompa-nied by the ejection of energetic protons whose (often very rapid) arrival in near-Earth space creates a hostile radiation environment. The resulting *solar radiation storms* are damaging to satellites, but can also be damaging to human tissue: fast protons—solar cosmic rays by any other name—can cause breaks in the double strands of DNA as they tear through. Such lesions can lead to mutation and tumorigenesis—astronauts are ill-advised to undertake extravehicular activities (spacewalks) during a solar radiation storm. During a particularly intense storm in 1989, cosmonauts aboard the Soviet *Mir* space station reportedly accumulated the maximum dose of radiation exposure for an entire year in a single day.

Spacecraft like Galileo, reaching Jupiter in 1995, or Cassini which arrived at Saturn in 2004, can be "hardened" against the deleterious effects of radiation during interplane-tary flight: astronauts cannot! Long-duration travel to Mars and back obviously carries a risk of occasional exposure to intense radiation conditions, and reliable shielding on board any manned spacecraft, coupled with accurate Space Weather forecasting, will be essential to the success of any such mission.

With increasing dependence on satellites for broadcasting, communication, en-vironmental and other monitoring, and navigation, Space Weather has become an issue that affects a considerable part of our daily lives. The auroral displays that we enjoy from time to time are, in many ways, a side-effect of the many and complex interactions between Earth's atmospheric envelope and the planet's extended space environment.

CHAPTER TWO

Causes of the Aurora

The aurora is a phenomenon of Earth's high atmosphere, occurring at altitudes typically in excess of 100 kilometers. Its light results from excitation of tenuous atmospheric oxygen and nitrogen by electrons accelerated in near-Earth space under conditions where the magnetosphere—the volume of space in which terrestrial magnetism has a dominant influence—becomes disturbed. Geographically, the aurora is ever-present in two oval regions, one surrounding either geomagnetic pole. These auroral ovals are normally quite narrow, and lie at high latitudes. Under disturbed conditions, the ovals brighten and broaden, and during severe disturbances—geomagnetic storms—they can be driven down toward the equator, bringing Nature's awesome light show to the skies of observers in the southern United States or more temperate parts of Europe. Other, less severe disturbances cause frequent substorms at higher latitudes: during these events, observers in Alaska or Canada, for example, will see the aurora brighten and become more active. Even at times when the Sun appears relatively inactive, variations in the solar magnetic field—particularly those associated with the so-called coronal holes—can trigger minor enhancements in auroral activity.

The disturbances that cause geomagnetic storms have their origins with solar activity, the most visible manifestations of which are the dark sunspots that come and go over a roughly 11-year cycle. Activity at, or just above, the Sun's surface influences magnetic conditions in interplanetary space, which is pervaded by the continuously outflowing solar wind. Since the dawn of the Space Age, scientists have come to better understand the nature of the solar wind and its influence on Earth (and the other planets). As a prelude to describing the aurora and how it can be observed and recorded, it is worth gaining an understanding of the Sun–Earth link and how this controls the levels and extent of auroral activity, as well as looking at the application of this understanding to the art of Space Weather forecasting.

Figure 2.1. Spectacular aurora; displays penetrating to lower latitudes, such as this event photographed from Edinburgh on 21–22 January 2005, are associated with violent solar activity. Our understanding of the underlying causes of such storms has improved immensely since the 1950s. Image: Dr Dave Gavine.

The Variable Sun

Our Sun is, like others, a variable star. Fortunately for the continuing existence of life on Earth, the Sun's variations in energy output are much less extreme than those that amateur astronomers can follow among countless stars in the night sky. Sunspots provide the most obvious indication of the Sun's variability, but there are other ways in which solar activity manifests itself and can influence conditions in interplanetary space.

Solar radiation is, of course, essential to life on our planet. Solar heating drives weather systems and ocean currents, while the energy of sunlight is harnessed in the process of photosynthesis by green plants forming the basis of food chains on Earth and—as a by-product—releasing oxygen into the atmosphere. Astronomers refer to the amount of radiation from the Sun arriving at the top of Earth's atmosphere as the Solar Constant, the average value of which from 1978 to 1998 was 1366.2 ± 1.0 watts per square meter. As the error bar on this value indicates, measurements from orbiting satellites reveal that this is, in fact, anything but constant, varying by about 0.1% over the course of the sunspot cycle. Ultraviolet and X-ray emissions from spotgroups and their associated bright faculae result in higher values for the Solar Constant at the maximum of the sunspot cycle. It has been estimated that this variation could give rise to global temperature change of the order of 0.1 K: the contribution of

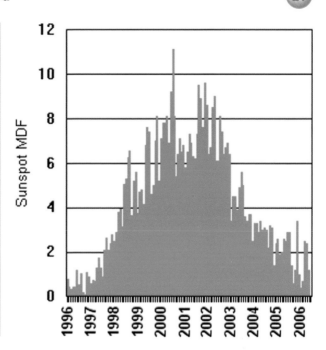

Figure 2.2.
Sunspot activity in cycle 23, expressed in terms of Mean Daily Frequency of active areas, as explained in Chapter 3. The cycle showed a double peak, with maxima in 2000 and 2001.

human industrial activities to global warming, is, however, a far more significant influence.

Past episodes of greater solar variability could have had a greater influence on global climate. For a long period, from about 1645 to 1715 (popularly known as the Maunder Minimum), sunspots appear to have been virtually absent—as were auroral displays at lower latitudes. Attention was first drawn to this by Gustav Sporer (1822–1895). Later, E. W. Maunder (1851–1928) of the Royal Greenwich Observatory renewed interest in the possibility of the interlude of diminished solar and auroral activity that now bears his name. Jack Eddy, a solar astronomer at the High Altitude Observatory, Boulder, Colorado subsequently brought the idea to major prominence during the 1970s.

From soon after the telescope's first application to astronomical purposes in 1609, until about 1645, sunspots were regularly seen and recorded by observers. Thereafter, they were seen only very infrequently. John Flamsteed (1646–1719), the first English Astronomer Royal, writing in 1684 reports:

"These appearances, however frequent in the days of Scheiner and Galileo, have been so rare of late that this is the only one I have seen in his face since December 1676."

Chinese astronomers, who made many naked eye sightings of sunspots in pre-telescopic times, similarly recorded few such observations in the period from 1639–1720.

At the same time, the records indicate a corresponding dearth of aurorae, from which researchers into long-term solar activity conclude that the shortage of sunspots was genuine, and not merely a result of lack of observer interest. The extensive catalog of historical auroral sightings (*Verzeichniss Beobachteter Polarlichter*) published by the German astronomer Hermann Fritz (1830–1883) for example, lists only 77 European aurorae during this period, a quarter of them from 1707–1708, when sunspot activity may have begun to revive. Many of the reports come from higher latitudes.

Data from other sources, such as the deposition of ^{14}C in annually laid-down tree rings, also indicate that the Sun was less active during a 70 year spell between the sixteenth and seventeenth centuries. The radioactive isotope ^{14}C ("carbon fourteen") is a natural product of the interaction between high-energy (Galactic) cosmic rays and nitrogen in the Earth's atmosphere. At times of high solar activity—accompanied by a strengthening of the general magnetic field in the solar vicinity—fewer of these extrasolar cosmic rays can penetrate the inner Solar System, and less ^{14}C is therefore produced in the atmosphere.

Atmospheric ^{14}C (from carbon dioxide) becomes assimilated into the newly growing parts of plants. Sections can be taken through long-lived trees, such as the California redwoods, and amounts of ^{14}C incorporated into annual growth rings measured in the laboratory. Results from several independent sources around the world do, indeed, suggest that the late seventeenth to early eighteenth century was a sustained period of high atmospheric ^{14}C abundance, consistent with diminished solar activity.

During the Maunder Minimum, global temperatures dropped by 0.5 K, perhaps corresponding with a 0.2–0.5% decrease in the Solar Constant. The period was also marked by low winter temperatures in the northern hemisphere, resulting in the final collapse of the Norwegian colony in Greenland. The frozen River Thames in London was the venue of winter Ice Fairs.

Direct observation of the equivalents of sunspots on other stars is impossible. However, observations at the light wavelengths emitted by ionized calcium have been used to monitor activity in the chromospheres of Sun-like stars, and this can be correlated with brightness variations. It appears that some of these stars may spend as much as one-third of their time in a "Maunder Minimum" low-activity state.

With some minor deviations (mainly variations in peak intensity), our Sun has been relatively well-behaved in the almost 300 years since the end of the Maunder Minimum, with the roughly 11-year sunspot cycle performing reasonably regularly.

The Maunder Minimum may have been only one of several such "excursions" in solar/auroral activity in historical times. Pre-telescopic records of naked eye sunspot sightings (made, for example, by observers in the Far East looking at the low Sun through haze—a practice not to be followed in the interests of protecting eyesight!) and, indeed, of aurorae, have been used to infer solar activity—both unusually high and unusually low—in the more distant past. Some of the conclusions can be backed up by ^{14}C tree-ring records (dendrochronology), while Chinese sunspot records are considered reliable back to the time of the Han Dynasty in 200 BC.

Some researchers consider it possible to detect interludes of high and low sunspot activity in the Chinese annals. Evidence is found, for example, of high sunspot activity in the twelfth century, a period which provided a rich crop of sightings recorded in monastic writings from Scotland and England. Conversely, sunspot activity was apparently low from the late sixth to early ninth centuries, and from 1403 to 1520 AD.

The putative gap in sunspot activity in the fifteenth and sixteenth centuries coincides with a period of high atmospheric ^{14}C concentration, and has been termed the Sporer Minimum. Low atmospheric ^{14}C concentration, and frequent sightings of both aurorae and naked eye sunspots lend support to the possibility of a Medieval Maximum around the twelfth century.

Several other maxima and minima—some of them more severe in their deviation from the present-day norm than the Maunder Minimum—may be postulated as far back as the Bronze Age on the basis of ^{14}C abundance studies, but it is much more difficult to find auroral, sunspot, or other supporting observational evidence. Jack Eddy has suggested the possibility of long-term cycles of solar activity (spanning as much as 1000 years each), in which excursions such as the proposed Medieval Maximum and Maunder Minimum represent peaks and troughs, respectively. It is possible, if Eddy's assumptions are correct, that activity is currently on a gradual rise toward a "supermaximum" in the 22nd and 23rd centuries. Conversely, a dendrochronological study by scientists at Germany's Max Planck Institut suggests that sunspot numbers in recent maxima have been at their highest for 6000 years, but that there is only a 1% chance of this situation persisting as late as the end of the 21st century.

Extending the record still farther back in time, sedimentary deposits —varves—laid down in lakes show periodic structural features which indicate that the cycle of rising and falling sunspot activity has been present for many millions, probably hundreds of millions, of years.

Many attempts have been made to correlate weather with sunspot activity, and there are possibly some direct links, to which we shall return at the end of this chapter.

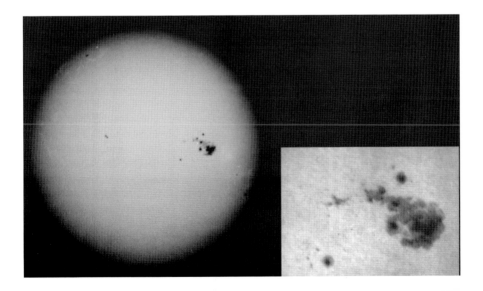

Figure 2.3. Full-disk photograph of the Sun on 26 March 1991 showing AR 6555. This giant sunspot group, shown in detail in the inset at right, was the source of major geomagnetic activity and extensive auroral displays. Image: Bruce Hardie.

The Sun as a Star

The Sun has a diameter of 1,400,000 km, and is considered to be a fairly normal Main Sequence G-class star about midway through its 10 billion year lifespan. Astronomers are able to infer the Sun's internal structure from an understanding of the processes by which it shines, and also more recently from the study of soundwaves traveling through its outer layers, the field of helioseismology.

The innermost third of the Sun comprises its superdense, extremely high temperature *core*. Here, at temperatures of 15 million K, nuclear fusion of hydrogen to generate helium occurs continuously. These nuclear reactions, predominantly the proton–proton reaction, are the source of solar light and heat. Photons of light emerge after long intervals following their "random walk" outwards through the overlying layers. Owing to collisions with particles in the overlying *radiative zone*, it is believed that each photon produced in the core of the Sun takes some 10,000 to 100,000 years to finally emerge at the surface. Little bulk motion is thought to occur in the radiative zone, but the Sun's outer third—the *convective zone*—extending from the deep interior to the visible surface of the *photosphere* is constantly turning over.

Convection on several scales is apparent in the Sun. Most obvious are the small "rice-grain" convective cells, observable under good conditions through even quite small Earth-based telescopes as granulation in the brightly shining photosphere. Small convection cells are on the order of 1000 km to a side, and have fairly short lifespans—of the order of 10 minutes. Larger convection cells in the photosphere— the supergranulation—tend to be longer-lived, persisting for perhaps 24 hours. The boundaries of supergranules, which may be 30,000 km to a side, are frequently marked by spicules and other indicators of intense magnetic field activity.

The outermost layer of the Sun shows *differential rotation*. The relative motions of sunspot groups at different solar latitudes suggest a rotation period of about 25 days for the solar equator, and as long as 36 days for the poles. Helioseismology shows differential rotation to affect the convective layer to a depth of 200,000 km, below which level, the Sun rotates as a solid body. The strong magnetic field of the Sun appears to be a consequence of differential rotation, coupled with convective processes.

Immediately overlying the bright visible surface of the photosphere, the innermost part of the Sun's extended atmosphere is the *chromosphere* ("color layer"), a region 10,000 km deep. The chromosphere becomes briefly visible during total solar eclipses as a ring of red light surrounding the dark body of the Moon. Its color is produced by excited hydrogen, emitting at a wavelength of 656.3 nm. Filters and spectrohelioscopes have been developed which allow astronomers to study the Sun in this restricted wavelength, the hydrogen-alpha line. Such observations have revealed much about activity in the chromosphere.

Chromospheric temperatures are higher than those of the photosphere, of the order of 10,000 K. A sharp temperature transition is seen between the two layers. Heating of the chromosphere probably results from upward-traveling shock waves from the photosphere, and also downwards transfer of heat from the inner corona.

The *corona*, the extended outer region of the Sun's atmosphere overlying the chromosphere, is the single most dramatic feature visible during a total solar eclipse. For a long time, the corona was held to be a relatively static, unchanging part of the Sun,

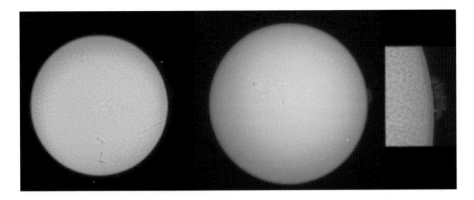

Figure 2.4. Using modern equipment, amateur astronomers an successfully image the solar chromosphere. At left, (14 April 2006), several dark filaments (prominences in silhouette) can be seen. The right-hand image (4 April 2006) shows a prominence on the limb, enlarged in the inset. Taken with a Meade ETX 90 telescope equipped with a Thousand Oaks hydrogen-alpha objective filter. Images: Bill Leslie.

Figure 2.5. The solar corona, seen during a total eclipse close to sunspot maximum. At this time, the corona is evenly-distributed around the Sun. Image: NASA/JPL.

Figure 2.6. At sunspot minimum, as in the total solar eclipse of 29 March 2006, the corona becomes drawn out into long equatorial streamers. Image: Neil Bone.

altering its shape in response to the sunspot cycle but doing little else. At sunspot maximum, the corona is fairly evenly distributed around the Sun, while at sunspot minimum, it appears drawn out into long equatorial streamers, and is absent from the Sun's polar regions.

Gas in the corona is at extremely high temperatures, between 1 to 5 million K. Under these conditions, the constituent gas is completely ionized and is in a state most accurately described as a *plasma*. Plasma is highly electrically conductive. The greater part of the corona is comprised of protons (the positively charge nuclei of hydrogen atoms) and electrons, with an admixture of heavier ions. The question of how the corona attains such high temperatures seems to have been resolved by observations using the TRACE satellite in 1998, which indicated that loops of magnetic field in a sheet-like "magic carpet" structure rising from the chromosphere below were sufficient to transfer the required energy.

The differing configuration of the corona from sunspot maximum to sunspot minimum reflects the changing nature of solar magnetic fields. At sunspot maximum, the tenuous coronal plasma is largely confined by closed loops of magnetic fields (having both ends of their field lines embedded in the Sun) above active regions. By sunspot minimum, "open" magnetic fields with one end embedded in the Sun and the other extending outwards to interplanetary space come to predominate, and coronal material spreads out more readily.

Two principal types of structure can be observed in the corona at visible light wavelengths: streamers and "helmet" structures. The bright helmets are produced by plasma trapped under the closed magnetic field loops above active regions. Streamers, on the other hand, emerge more or less radially from regions where solar magnetic field lines are open. Long streamers are thus more frequently present and obvious in the corona around sunspot minimum.

The inner parts of the Sun's corona can be routinely studied from high-altitude ground-based observatories by use of the Lyot coronagraph, which provides, optically, an artificial eclipse within the telescope. Ground observers, however, still obtain their most detailed views of the outer corona during rare total solar eclipses. On such occasions, scattering by Earth's atmosphere of sunlight from outside the zone of totality still results in a fairly bright sky background against which it is difficult to resolve fine-contrast features in the outer corona. The best observations of the corona are thus obtained using coronagraphs aboard orbiting satellites. The NASA/ESA SOHO (Solar and Heliospheric Observatory) spacecraft, in a "halo" orbit around the gravitationally stable L1 point between the Earth and Sun (1,500,000 km sunwards of Earth) has, since 1995, provided astonishing insights into the dynamic nature of the corona.

Satellite observatories have the advantage of being able to exploit emissions in the far ultraviolet and X-ray wavelengths (absorbed by Earth's atmosphere and therefore inaccessible to ground-based observers), which allow the corona to be traced to even greater distances from the Sun. X-ray wavelength observations made during the Skylab missions in 1973 and 1974 greatly advanced understanding of solar coronal processes, and laid the foundations for subsequent studies with the Solar Maximum Mission (SMM; "Solar Max") and Yohkoh satellites.

The corona is a strong emitter of X-rays as a consequence of its high temperature. The visible corona seen during total eclipses is produced by the scattering of sunlight by electrons in the inner solar atmosphere. Heavier atomic species, such as iron, also become ionized, and their emissions at X-ray wavelengths may be used to trace the corona to great distances outwards from the solar disk. Some coronal streamers may reach outwards for as much as 100 solar radii, halfway to the Earth.

Sunspots and the Solar Magnetic Field

The Sun's magnetic field arises in its outer third, and is amplified and brought to the surface by convection. Movements of the highly conductive solar gas are determined by the magnetic field: gas is constrained to flow along, but not across, solar magnetic field lines. As a result, the smooth pattern of the convective granules can be disrupted in regions where magnetic flux emerges from the solar interior through the photosphere. Gas rising to the surface in such regions is restrained by magnetic fields whose strength may be of the order of 1000–3000 Gauss (for comparison, Earth's magnetic field has an intensity of 0.3–0.6 Gauss). Prevented from moving laterally and sinking again, the gas cools, resulting in the appearance of sunspots.

Sunspots appear dark against the surrounding, hotter photosphere. Typical sunspot temperatures are of the order of 4000 K, compared with 6000 K for their immediate surroundings. They are therefore not particularly cool by terrestrial standards: if a large sunspot could be isolated from the photosphere and suspended in the night sky, it would glow reddish and provide much more illumination than the Full Moon! The photosphere itself is fairly shallow in comparison with the convective layer as a whole. The visible light of the Sun comes from the top 100 km or so, levels much below this being opaque to light.

Sunspots show wide variations of size, appearance, structure, and duration. The smallest sunspots are simple pores, lasting only a few hours. More complex groups may begin life as pores, growing in size to affect large areas of the solar disk. Large sunspots show two distinct regions, a darker (and cooler) central umbra, surrounded by a lighter penumbra. Structures in the penumbra show alignment to strong local magnetic fields. Complex spot groups may contain several umbrae embedded in a mass of penumbra, and surrounded by pores. The appearance of such groups, which can become sufficiently large to be visible to the suitably protected naked eye, may change from hour to hour: such rapidly changing, complex spot groups are often the source of solar flares.

The largest sunspot groups may come to cover quite significant areas of the solar disk. Apparent sunspot areas may be estimated using calibrated projection disks (simple solar observation methods are described in Chapter 3). Extensive groups, such as that involved in producing the Great Aurora in March 1989, may cover as much as 3600 millionths (0.4%) of a solar hemisphere: by comparison, the Earth is of insignificant size.

In addition to varying in number over the roughly 11-year cycle, sunspots undergo a regular change in magnetic field orientation. The magnetic fields associated with sunspots were first observed by George Ellery Hale (1868–1938) and his colleagues at Mt. Wilson in 1908, who found that leader and follower spots within a group show opposite magnetic polarities. Magnetic polarities are reversed, however, for leader and follower spots in the other hemisphere of the Sun at the same time. The picture is further complicated by a reversal of leader and follower spot magnetic polarities in either hemisphere at the end of each approximate 11-year cycle: the complete magnetic cycle takes about 22 years to return to its starting configuration.

The most widely accepted model for the production of sunspots is that proposed by Horace W. Babcock in the early 1960s, and subsequently developed by Robert B. Leighton. This model accounts both for the appearance of sunspots, and the reversal of the solar magnetic field that accompanies each new sunspot cycle.

Solar magnetic field lines may be initially visualized as lying along meridians running north–south. Differential rotation in the outer layers soon begins to stretch field lines horizontally, concentrating them. Where several field lines are brought into close proximity, loops of magnetic flux are forced to the surface, breaking through the photosphere where they disrupt convection and give rise to spot groups.

Field loops initially break through at higher heliographic latitudes. The first spots of a cycle therefore emerge at higher solar latitudes. The zones of latitude at which spots appear most frequently gradually migrate equatorwards. Plotted against time, the distribution of spot latitudes gives rise to the Maunder "butterfly diagram." Cycles overlap, so that the first, high-latitude, spots of a new 11-year cycle begin to appear while the last near-equatorial spots of the previous cycle are present.

The reversal of polarity in sunspot groups from one cycle to the next is accounted for by the gradual dispersal of magnetic flux from decaying active regions. As a consequence of the winding-up of magnetic field lines by differential rotation, leader and follower (eastern and western in longitude within a group) spots appear at slightly different latitudes: the leader is usually closer to the equator. As an active region decays and its magnetic flux diffuses outwards, more of the leader's magnetic flux will cross the equator into the other solar hemisphere. Consequently, toward the end of one cycle and the beginning of the next, each solar hemisphere comes to acquire a net magnetic polarity derived from that of the leader spots in the opposite hemisphere during the cycle immediately past. When each new cycle starts, magnetic polarities therefore begin from a reversed configuration.

Associated with the regions in which sunspots appear are *faculae*, observed in white light as brighter areas of the photosphere. Faculae are seen to best advantage when close to the limb of the projected disk, where they appear brighter in contrast against the limb darkening (a consequence of the absorption of light from the photosphere by its passage through a greater volume of solar atmosphere in line of sight to the observer than at the centre of the disk). They persist for some months, often appearing before sunspot activity manifests itself at that position on the disk, and remaining for some time after the associated sunspots have decayed.

Faculae lie in the upper reaches of the photosphere, corresponding roughly with the plages observed in hydrogen-alpha wavelengths in the overlying chromosphere. These appear to be regions of enhanced gas density and temperature.

The active regions in which sunspots are involved are important in generating aurorae. The most vigorous aurorae are generally those that follow the ejection, from the neighborhood of sunspot groups, of energetic particles during solar flares. The large, actively changing sunspots with which such flares are most frequently associated are usually commonest in the run-up to sunspot maximum, and this, too, is the time when aurorae might most often be expected to extend toward mid-latitudes.

The Active Chromosphere and Corona

The photosphere is the most accessible region of the Sun for white light observation, and sunspots are the most obvious indication of magnetically disturbed conditions near the solar surface. Sunspots, however, are only part of the extensive and complex combination of effects that comprise active regions on the Sun. Activity in the chromosphere and corona overlying sunspot groups is of great importance with respect to solar-terrestrial relations.

Solar prominences extending a short distance outwards from the Sun's limb may be a striking feature during total eclipses. Prominences consist of relatively cool gas from the chromosphere, or condensed from the corona above, suspended in the inner solar atmosphere by magnetic fields. The gas in prominences remains cool as a result of the insulating effect of the associated magnetic fields: conduction of heat from the surrounding, hotter corona is inhibited.

The use of hydrogen-alpha filters allows prominence activity to be routinely monitored on a day-to-day basis. Prominences are seen to best advantage when presented on the limb, reaching heights of as much as 50,000 km (getting on for a solar radius) above the photosphere. They may also be visualized when in transit across the disk as *filaments* in hydrogen-alpha, appearing dark in contrast against the brighter background of the Sun. Prominences are relatively long-lived features, and may persist for up to four or five months.

Prominences occasionally erupt from the chromosphere outwards into the corona. These events are associated with coronal transients, or mass ejections. The eruption of prominences while in transit across the Sun's Earth-facing hemisphere may occasionally be observed as a "disappearing filament" events. These sometimes unload particles Earthward into interplanetary space and may be followed a day or so later by

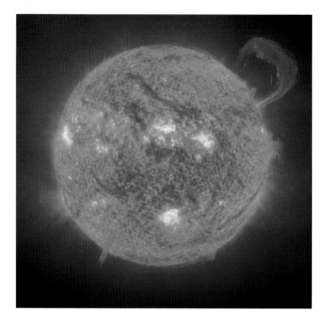

Figure 2.7.
SOHO image
showing a giant
solar prominence
and bright active
rgions (plages) in the
chromosphere on
14 September
1998. Image: ESA.

enhanced geomagnetic activity and aurorae at lower latitudes. In two-thirds of cases, the prominence becomes re-established within a couple of days of a disappearance event.

When observed as filaments against the solar disk, prominences are seen to lie along the boundary between areas of opposite magnetic polarity in active regions, the neutral line. Hydrogen-alpha spectroheliograms show active regions outlined as bright areas—*plages*—against the solar disk.

Prominence activity is often, but not always, associated with sunspot regions. Prominences commonly appear in decaying active regions once the sunspots have dissipated. The frequency with which quiescent prominences are seen at equatorial and tropical heliographic latitudes roughly follows the sunspot cycle, peaking a couple of years later than the spot groups. Active prominences, produced by condensation of material falling back from the corona following violent ejections, are seen at times of high solar activity.

The corona as studied from space is seen to be far from static. Mass ejection events associated with solar flares or the disappearance of prominences lead to the appearance of denser "bubbles"—referred to in the 1970s as transients, now more commonly as coronal mass ejections (CMEs)—traveling outwards through the corona. Particularly around the time of sunspot maximum, the inner corona is subject to quite rapid turnover: transients may remove the equivalent of the whole mass of the corona in three months at sunspot maximum. Following its removal during a CME coronal plasma is quickly replenished from below.

X-ray observations from orbit have been of particular value in confirming the occurrence of several predicted phenomena in the corona. The photosphere appears dark (being too cool to emit at such wavelengths) when the Sun is observed in X-rays. Such observations therefore allow the corona to be observed in its entirety.

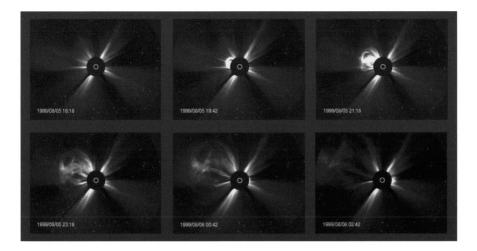

Figure 2.8. Progress of a Coronal Mass Ejection (CME) observed over an eight-hour period on 5–6 August 1999 with the LASCO C3 coronagraph aboard SOHO. Image: ESA.

The X-ray corona in the direction of the solar disk is seen to contain a number of features. Active regions are revealed as bright in X-rays. Numerous bright points, interpreted as the bases of coronal streamers are seen. Several short-lived bright spots, with durations of the order of eight hours, are also seen against the solar disk: these may give rise to small, but frequent, releases of magnetic flux from the Sun and may be important in heating the corona.

Coronal Holes

Of particular interest are the coronal holes, first well visualized from Skylab in the early 1970s. In regions where the magnetic field lines emerging from the Sun are "open," coronal gas is no longer confined, and can flow outwards and cool relatively freely. These regions therefore have lower gas densities and temperatures than their surroundings, and appear darker in X-rays. These streams of coronal gas sweep outwards into interplanetary space, where they may be encountered by the Earth at regular intervals corresponding to the Sun's rotation. Coronal hole streams can persist for many months, giving rise to recurrent series of quiet aurorae over the period of their existence.

The polar regions of the Sun are thought to be occupied by permanent coronal holes. Seasonal variations in the occurrence of quiet aurorae can be accounted for by the differing presentation of the Earth to the polar coronal hole streams as it orbits: around the equinoxes, the inclination of the Earth's orbit relative to the solar equator brings it more directly into line with the Sun's higher latitudes.

Coronal holes extending to the Sun's equatorial regions tend to appear toward the end of the sunspot cycle, and are the major source of highly magnetized particle

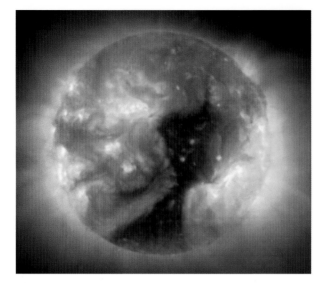

Figure 2.9.
SOHO EIT image of
a coronal hole—
appearing dark in
the lower centre of
the solar disk—on 8
October 2002.
Image: ESA.

streams (and consequent enhancements of terrestrial auroral activity) at this time. The equatorial coronal holes eventually become connected to the polar holes, and gradually retreat polewards as they slowly decay.

Solar Flares

While coronal holes are the major source of disturbed geomagnetic conditions in the years leading up to sunspot minimum, by far the most vigorous disturbances result from violent activity in the inner atmosphere of the Sun at those times when sunspots are common. The association between high sunspot numbers and the enhanced likelihood of extensive auroral storms has long been known. Observation by the English astronomer Richard Carrington of a white light solar flare in a sunspot group on 1 September 1859, followed a day or so later by major auroral activity, began to cement the connection.

Solar flares occur not within the sunspot groups themselves, however, but in the inner corona immediately above active regions. In these locations, intense contorted magnetic field lines are brought into close proximity. As with many solar phenomena, precise mechanisms are still being investigated, but it is widely accepted that solar flares begin from a small, intense "kernel," within which vast energies are released over short time intervals as a result of magnetic reconnection (coming together of regions of opposed magnetic polarity), and from which shock-waves propagate outwards.

Shock-waves propagating downwards from the kernel to the photosphere give rise to a characteristic two-ribbon structure in major flares. Visualized in hydrogen-alpha light, this often begins life as a series of bright points in an active region, joining to become continuous. Two-ribbon flares apparently mark the footpoints of "arcades" of material either side of a magnetically neutral zone.

Figure 2.10.
Bright solar flare in October 2003, recorded by the SOHO EIT. Image: ESA.

Prominences up to 100,000 km away from the site of a large flare may be disturbed by outward-propagating shock waves traveling through the inner solar atmosphere. Particles accelerated to high energies within flare kernels are ejected outwards into the corona.

The onset of a solar flare is marked by an abrupt increase in the X-ray and ultraviolet wavelength emissions from the region in which it is to occur. Characteristically, the rise of a flare to its peak of intensity is more rapid than the subsequent decline. Flares may have durations as short as 10 minutes, or may persist for several hours.

Energetic electrons ejected during solar flares produce bursts of radio noise as they collide with coronal material. Type II radio bursts are detected between 10 and 30 minutes following the onset of a flare, and result from shock-waves traveling outwards at velocities of 800–2000 km/s through the corona. Electrons accelerated to higher velocities—perhaps as much as half the speed of light—give rise to type III radio bursts. Most energetic are type IV radio bursts, which sometimes follow type II, and are caused by pockets of magnetized plasma traveling out through the corona; these appear to be associated with coronal transients.

Flares vary considerably in magnitude (Tables 2.1 and 2.2). The most intense solar flares are associated with the most complex active regions, which are often also those containing the most extensive and complex sunspot groups. Such groups are perhaps commonest during the early rise of the sunspot cycle towards maximum.

Around sunspot maximum, there may be as many as 25 flares per day on the observable hemisphere of the Sun.

Solar flares are only very rarely visible in white light; most are visualized by routine monitoring of the Sun in hydrogen-alpha wavelengths from ground observatories, or in ultraviolet and X-ray wavelengths from orbiting satellites. As yet, it remains impossible to predict such events more than an hour or so in advance. The onset of a flare is often preceded by disturbances in the magnetic pattern above an active region,

Table 2.1. Intensities of Solar Flares in H-alpha

Type	Area of solar hemisphere affected (millionths)
Sub-flare(s)	<100
1	100–250
2	250–600
3	600–1200
4	>1200

Flares of each type may be further classified:
  F Faint
  N Normal
  B Bright

Therefore, the most intense events will be Type 4B

but by the time such disturbances are noticeable, the flare itself is often already in progress.

Linking the Sun and the Earth—The Solar Wind

Plasma in the Sun's corona is largely trapped under magnetic field loops lying above active regions; indeed, without such constraining fields, coronal material would never be sufficiently concentrated to become visible. Solar magnetic field loops reaching to great distances above the photosphere may eventually become so stretched that they become open, however, allowing coronal plasma to flow freely into interplanetary space: essentially, the kinetic energy of the plasma exceeds the restraining energy of the local magnetic field in such regions. The corona appears to contain several such open regions at any one time, from which long streamers emerge past the closed regions delineated by helmet structures, fanning out as they extend farther from the Sun.

Open regions in the Sun's magnetic field are the source of the *solar wind*, a continual outflow of plasma from the Sun that permeates the entire Solar System. Confirmation of the existence of the predicted solar wind, postulated by Eugene N. Parker and others

Table 2.2. Intensities of Solar Flares in X-rays

Importance	Peak flux (watts/m^2)
A	10^{-8} to 10^{-7}
B	10^{-7} to 10^{-6}
C	10^{-6} to 10^{-5}
M	10^{-5} to 10^{-4}
X	10^{-4} and greater

into the late 1950s, was one of the early landmark successes of planetary exploration by spacecraft. Equipment aboard Mariner 2, launched toward Venus in 1962, detected the solar wind flowing outwards at 400 km/s.

The solar wind is not always a steady, quiescent flow. The violent ejections of material associated with large solar flares or the disappearances of prominences lead to increased densities and speeds. Solar flares may introduce turbulent pockets of high solar wind velocities reaching 1000–1200 km/s. Particle streams from coronal holes produce local solar wind velocities of 800 km/s.

Solar wind plasma consists principally of protons and electrons, with a small population of other atomic nuclear species. The plasma has, frozen into it, a magnetic field, whose strength and orientation are determined by features at the Sun's surface. The strength and direction of this Interplanetary Magnetic Field (IMF) carried along with the solar wind is important in determining the nature and extent of its interactions with the Earth's magnetosphere (the volume of space in which the terrestrial magnetic field has a dominant influence over particle motions). In particular, when the IMF has a strong southerly component relative to the ecliptic plane, conditions are favorable for the occurrence of more active and extensive aurorae.

Emerging plasma initially co-rotates with the Sun below. The solar wind velocity increases with increasing distance from the Sun, as the thermal energy of the plasma overcomes magnetic and gravitational restraints. From five to six solar radii (3,500,000–4,200,000 km) outwards, the solar wind becomes supersonic. Close to the Sun, but above closed coronal loops, "open" field lines emerge radially from the Sun. As these field lines are dragged outwards by the solar wind, the rotation of the Sun which carries their points of emergence, begins to wrap them into spirals—the so-called "Parker spiral" pattern. The Interplanetary Magnetic Field becomes increasingly wound into a spiral structure with distance from the Sun. Viewed from above the plane of the ecliptic, the solar wind magnetic field makes an angle of about 45° relative to the direct Sun–Earth line at the distance of the Earth's orbit. The IMF passing the Earth corresponds to the magnetic field of features near the central meridian of the observed solar disk 36–48 hours previously.

Fine structure in the IMF carried by the solar wind has an important influence on the nature and extent of terrestrial auroral activity and, indeed, that observed in the atmospheres of other planets. On a larger scale, the Sun appears to have semi-permanent regions of magnetic polarity in either hemisphere. These large-scale regions are also evident in the solar wind as *magnetic sectors* in the plane of the solar equator.

There are usually two or four of these sectors, more or less evenly spaced. Between the hemispheres in the solar wind lies a neutral sheet, roughly in the plane of the ecliptic. Solar wind velocities are lowest in the neutral sheet, rising sharply with increasing ecliptic latitude. Measurements obtained using instruments aboard the Ulysses spacecraft (Chapter 7) in 1994 also indicated a marked increase in solar wind speed at higher heliographic latitudes. The solar wind emerging from the permanent polar coronal holes has a velocity of 750–800 km/s.

Around sunspot minimum, the neutral sheet in the solar wind is fairly flat. At times of higher activity, however, it becomes folded into a "pleated" form, the folds appearing steeper relative to the ecliptic plane at sunspot maximum. The folds of the neutral sheet are, of course, the borders between regions of opposed magnetic polarity in the solar wind. During each solar rotation, folds in the pleated sheet sweep across the Earth from time to time, leading to fairly abrupt changes in the local IMF. These *sector boundary crossings* can lead to short-term increases in geomagnetic activity.

The neutral sheet appears to be largely confined to the lower solar latitudes. Pioneer 11, which passed Jupiter in 1974, was gravitationally slung slightly above the ecliptic plane en route to its 1979 encounter with Saturn. During its out-of-ecliptic passage, Pioneer 11 experienced long periods of constant solar wind magnetic polarity, implying that the changes associated with sector boundary crossings and the neutral sheet are indeed principally a phenomenon associated with equatorial heliographic latitudes.

New Insights on Coronal Mass Ejections and Solar Flares

During the early 1990s, an attractive alternative to the "flare paradigm" was developed, notably by Jack Gosling of Los Alamos National Laboratory and Art Hundhausen of the High Altitude Observatory at Boulder. Gosling and Hundhausen cite coronal mass ejections rather than flares as the main class of solar activity responsible for causing major geomagnetic disturbances and low-latitude aurorae.

Modeling of CMEs by Hundhausen suggests that these develop from helmet structures in the corona which become detached from their chromospheric footpoints. Subsequent reorganization of the underlying magnetic field may result (20–30 minutes later) in a flare event: in a reversal of previous dogma, flares may be a consequence of magnetic field reconnection following the CME, rather than vice versa!

Measurements from the ISEE-3 satellite, stationed sunwards of the Earth around the L1 stable orbital point from 1978 to 1982, showed counterstreaming electron currents, characteristic of CMEs, ahead of all but one of the 14 most vigorous magnetic storms in this interval. In the next sunspot cycle, 36 out of the 37 biggest geomagnetic storms were associated with CMEs. Observations at X-ray wavelengths from the Yohkoh satellite in 1992 lend support to the view that CMEs can occur without solar flares as the trigger.

Traveling at velocities which can occasionally exceed 1000 km/s, well in excess of the undisturbed solar wind, these ejections pile into material ahead of them, creating an interplanetary shock wave. CMEs often show a dark trailing edge as they sweep up the slower-moving solar wind in front. The arrival of the shock wave can severely distort Earth's magnetosphere, resulting in geomagnetic storms and their associated lower latitude auroral activity. Furthermore, the "draping" of field lines ahead of the shock wave may help to satisfy the requirement for a southerly configuration in the interplanetary magnetic field to trigger a major geomagnetic disturbance.

The ACE spacecraft, stationed around the L1 Lagrangian point 1.5 million km sunwards from Earth in the solar wind now affords early warning of the arrival of CME-associated shockwaves and magnetic fields, detecting these an hour or so before they reach the planet.

Interactions with the Solar Wind

Comets

The most obvious interactions between the solar wind and small Solar System bodies are probably those observed in comets. Comet-solar wind interactions show some parallels with those between planetary magnetospheres and the solar wind.

The Giotto spacecraft encounter with Comet 1P/Halley on 13 March 1986 broadly confirmed the "dirty snowball" cometary model proposed in the 1950s by Fred Whipple. The nucleus of 1P/Halley is an irregular body comprising dusty material bound up in an icy matrix, with a thin outer crust. Other comet nuclei that have been visited by spacecraft (19P/Borrelly, 81P/Wild 2, 9P/Tempel1) are similar, but subtly different. Debate continues as to whether these bodies are solid, or loosely held "rubble pile" aggregates. At great distances from the Sun, typically beyond the orbit of Saturn, comet nuclei are inert, and essentially indistinguishable from asteroidal objects. Within about 3 AU from the Sun, sublimation of cometary gases occurs in response to heating by solar radiation.

Initially confined to a small coma immediately surrounding the nucleus, material ejected from the comet soon becomes drawn out into tails by the action of the solar wind and radiation pressure. Dusty material forms a curved tail, as the small debris falls away to pursue an independent orbit around the Sun as a meteor stream. Gas, ionized by the ultraviolet component of solar radiation, forms narrower, more or less straight tails generally pointing directly away from the Sun. Fluorescence at 420 nm wavelength by carbon monoxide (CO^+) ions released into the coma is particularly marked in some comets, and allows the bluish plasma (or ion) tails to be quite easily traced.

The accepted mechanism by which cometary ion tails are produced is essentially that proposed by Hannes Alfven in the late 1950s. Ions produced by the action of solar ultraviolet on the coma become trapped along field lines of the Interplanetary Magnetic Field in the passing solar wind. Consequently, the solar wind is decelerated in the comet's neighborhood. Close to the nucleus, where the cometary ion concentration is highest, the solar wind magnetic field is excluded, and a cavity is formed. Interplanetary Magnetic Field lines become draped over the cavity's boundary, which is referred to as the *contact surface*. During its encounter with 1P/Halley, Giotto detected a contact surface 3600–4500 km from the nucleus. Within this region, the ion signature recorded using particle detectors changed from a mixture of heavy cometary ions and solar wind protons, to that of cometary ions alone.

Lobes of opposed magnetic polarity are produced along the comet tails. Visible ion tails lie along the neutral current sheet between these lobes. In many respects, this structure is a useful analogy for that seen in planetary magnetospheres, including that of Earth.

Another feature shared by comets and planetary magnetospheres is the *bow shock*, produced "upwind" of the contact surface. Likened to the wave pushed ahead of a ship's bow ploughing through water, the bow shock is a region around which solar wind plasma flows without further interaction with the comet. Disturbances of the solar wind's smooth flow may extend for a considerable distance downwind of the comet.

Cometary bow shocks are less clearly defined than those produced by planetary magnetospheres. The distance from the nucleus of the bow shock is governed by the comet's rate of gas production. Measurements from the ICE spacecraft at 21P/Giacobini-Zinner in 1985, and Giotto and other probes at 1P/Halley in 1986 showed the 1P/Halley bow shock to lie 10 times further (about 1 million km) from the nucleus; 1P/Halley was estimated to be 30 times more productive of gas.

Following a period of "hibernation" after the P/Halley encounter, Giotto was targeted to a further comet, 26P/Grigg-Skjellerup, which it encountered on 10 July 1992. 26P/Grigg-Skjellerup is considered to be older, probably smaller, and certainly much less active than either 21P/Giacobini-Zinner or 1P/Halley, and much depleted in the ices involved in gas production. During the 26P/Grigg-Skjellerup encounter, Giotto ran across a bow shock around 20,000 km from the nucleus. The comet's gas production rate was probably a hundredth that of 1P/Halley.

The interactions between the ion tails of comets and the Interplanetary Magnetic Field carried by the solar wind are of particular interest. Changes in the intensity and direction of the IMF give rise to *disconnection events*, in which ion tails that had been developing in one direction may become sheared off, following which new ion tails reflecting the changed IMF direction in the comet's vicinity begin to grow. Disconnection events may result from the crossing of sector boundaries in the neutral sheet of the solar wind or from encounters with pockets of disturbed solar wind introduced by coronal mass ejections. Ion tail disconnection events involve magnetic field line reconnections, beginning upwind on the sunward side of the nucleus, and are fairly gradual, taking 12 hours or more to complete.

One of the finest comets in recent memory was 1996B2/Hyakutake. Discovered on 31 January 1996 by Japanese amateur astronomer Yuuji Hyakutake (1950–2002), the comet passed Earth at a distance of only 0.1 AU in late March of that year, when it became a prominent magnitude 0 naked-eye object showing a long, bright ion tail. Although its appearance coincided with the declining phase of the sunspot cycle, the solar wind was at this time sufficiently "gusty" to produce several well-observed disconnection events.

Some, observing under very dark skies, reported what were initially taken to be improbably large values for the length of Comet Hyakutake's ion tail around closest approach. Observations suggesting an angular span of up to 100° across the sky from the nucleus to the tail's end were dismissed at the time as overestimates, and geometrically impossible. Reconsideration of the factors controlling the ion tail, however, has led to acceptance of these results. Instead of being absolutely straight, the ion tails of comets are slightly curved—each follows an arc along a "Parker spiral" of magnetic field emerging from the Sun, and such a curved geometry can accommodate the reported extent of Hyakutake's tail in 1996. Measurements from the Ulysses spacecraft also confirmed the tail's great length.

Comets might be regarded as natural probes of regions of the outer solar atmosphere to which Earth-based observers have no other access, as was recognized by Biermann in the early 1950s. Whipple has given the apposite description of comets as "solar wind-socks."

Several disconnection events were also observed during the perihelion passage of Comet Halley in 1986, mostly in association with sector-boundary crossings. It has been proposed that solar activity was responsible for a remarkable brightening of 1P/Halley observed in February 1991, when the comet lay 14.3 AU from the Sun (far

Figure 2.11. Comet 1996B2/Hyakutake, showing a disconnection event in its bright ion tail on 25 March 1996. Image: Nick James.

beyond the orbit of Saturn) at a distance where the nucleus would be expected to be dormant. Shock waves associated with solar flares at the end of January 1991 may have arrived in the comet's vicinity some two weeks later, acting as a trigger for unusual activity. Even out at the farthest fringe of the Solar System, the Voyager spacecraft have detected shockwaves in the solar wind associated with violent activity that took place in the near-solar environment many months previously.

Earth's Magnetosphere

It is to the complex interactions between the terrestrial magnetic field, and that carried by the solar wind as it passes Earth, that we must look to understand how the aurora is generated.

Earth's layered—differentiated—interior structure has been inferred from seismographic studies, tracing the propagation of shockwaves produced during earthquakes. The outer continental crust (up to 50 km deep) floats above a deeper mantle, comprised of two principal layers separated by a transition zone in which magma originate. Deep at the planet's center lies the core, which has a diameter of roughly 7000 km—about 45% of Earth's overall diameter. Comprised of nickel–iron, the core is under immense pressure (around 3850 kbar) and is at a high temperature, ranging from approximately 3000 K in its outer parts to 6000 K in the center. The innermost third of the core is solid, and is overlain by a fluid region. Between the two regions of the core, there is probably a relatively thin transition layer.

Electrical currents arising from rapid convective motions in the fluid outer core are the source of terrestrial magnetism. These motions may be driven by small-scale variations in chemical composition or by the radioactive decay of heavy elements. Evidence of terrestrial magnetism is found in the oldest rocks (formed 3.5 billion years ago), implying that the Earth underwent differentiation to its broad current structure fairly rapidly after formation, with denser materials like iron sinking toward the center.

The magnetic field generated in the Earth's core has an equatorial strength of 0.3 Gauss, and is about twice this strength in the polar regions. Studies of paleomagnetism, the magnetic field imprint frozen into solidifying rocks at past epochs, indicate that the Earth's magnetic field undergoes relatively regular reversals in the long term. The precise locations of the magnetic poles also vary gradually in the short term: movement of the north geomagnetic pole has been invoked as one explanation for the higher frequency of auroral sightings at mid-latitudes in Europe around the twelfth century relative to the present day. Over timescales of centuries, the magnetic field as a whole has drifted westwards. The 2005 position of the north geomagnetic pole was 82.7°N, 114.4°W in the Arctic Ocean north of Canada, near Resolute Bay: observers at relatively low geographical latitudes in North America enjoy a higher frequency of auroral occurrence than their European colleagues at identical latitudes, by virtue of being closer to the geomagnetic pole.

Structure of the Magnetosphere

The magnetic field generated by currents in the Earth's core is extensive. In isolation, it would have a simple dipole configuration, similar to the familiar, symmetrical pattern produced by field-lines revealed around a schoolchild's bar magnet when iron filings are sprinkled onto a piece of paper laid over it. Interactions between the Earth's magnetic field and the solar wind result in distortion: while 90% of terrestrial magnetism may be represented by a simple, regular dipole, the outermost 10% of the field is pushed earthwards on the side facing the Sun, and dragged downwind into a long "tail" on the night-side. The comet-shaped "cavity" in the solar wind within which terrestrial magnetism is dominant is referred to as the *magnetosphere*.

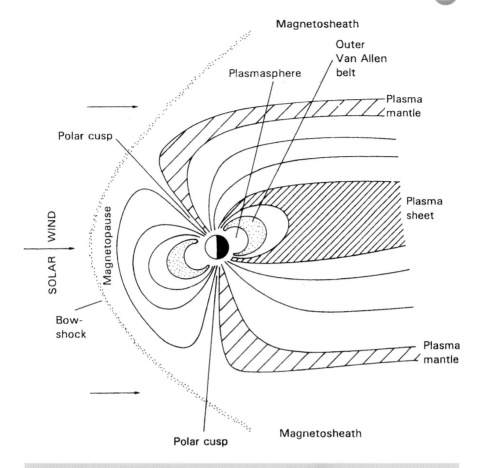

Figure 2.12. Schematic cross-section through the noon-midnight plane of the magnetosphere, illustrating how magnetic field lines ahead of Earth in the solar wind are compressed, while those in the anti-solar direction are drawn out downwind towards the magnetotail. Most of the solar wind plasma is deflected around the bow-shock, usually 64,000 km upwind of Earth. Some plasma, however, penetrates via the polar cusps. Plasma is carried downwind along the plasma mantle, and into the neutral sheet lying in the equatorial plane of the magnetotail. Reconnection between solar wind and terrestrial magnetic field lines at the leading edge of the magnetopause—the boundary of Earth's magnetic influence, which lies below the bow-shock—may result in ejection of plasma from the sheet. Plasma is ejected down the magnetotail back into the solar wind, and also into Earth's upper atmosphere, there producing the aurora.

While the structure of the magnetic field at ground level may be directly measured quite readily, an understanding of its precise behavior when extended into near-Earth space had to await the advent of satellite and space probe exploration starting in the late 1950s.

Earth's magnetic field is a barrier to the solar wind. About 10 Earth-radii (64,000 km) ahead of the Earth, a bow shock is produced in the solar wind as this encounters the leading edge of the terrestrial magnetic field. The bow-shock upstream of the Earth in solar wind is described as *collisionless*, resulting from field, rather than particle, interactions, and is a region where solar wind plasma becomes heated.

The bulk of the solar wind plasma flow is deflected the bow-shock and on into interplanetary space without any further interaction with the terrestrial magnetic field. The precise distance of the bow-shock changes in response to conditions in the solar wind: it moves earthwards when the solar wind pressure (velocity) is higher, and sunwards at times of low solar wind intensity—an almost continuous process of 'breathing' in and out.

The boundary of Earth's magnetic influence is delineated by the *magnetopause*, lying under the bow-shock. Between the bow-shock and magnetopause lies a plasma-rich region, the *magnetosheath*.

Terrestrial magnetic field lines in the direction facing toward the Sun are compressed, while those in the anti-solar direction are drawn out toward the distant *magnetotail*, which lies some millions of kilometers downstream in the solar wind, well beyond the Moon's orbital distance. Each month, for a number of days around full, the Moon passes through Earth's magnetotail, which has a cross-section of 40–60 Earth-radii (255,000–382,000 km).

Viewed in cross-section perpendicular to the Sun–Earth line, the magnetotail comprises two *lobes* of opposed magnetic polarity, derived from the separate hemispheres of the terrestrial magnetic field. Electric currents in the magnetosheath circulate in opposite senses around these lobes, meeting in the *neutral sheet* which lies in the equatorial plane of the magnetotail. The current in the neutral sheet flows eastward across the magnetotail from the dawn to the dusk side.

While the vast bulk passes the Earth without interaction, some solar wind plasma is able to funnel into the magnetosphere at higher latitudes via the polar cusps. An important route of entry of solar wind plasma is reconnection between Interplanetary Magnetic Field lines and the terrestrial magnetic field at the "nose" of the magnetopause. From here, it travels along the outside of the magnetosheath, then enters the *plasma sheet*, a region of hot plasma surrounding the neutral sheet. A much smaller amount of solar wind plasma also diffuses across the magnetosheath at lower latitudes.

Sunlight ionizes the rarefied upper atmosphere, producing the layers of the *ionosphere*. Charged particles from the ionosphere can "evaporate" upwards along magnetic field lines into the magnetosphere. Almost equal proportions of material in the neutral sheet may be derived from solar wind and terrestrial sources. Plasma resident in the neutral sheet can, under certain conditions, be injected into the Earth's high atmosphere, giving rise to enhanced auroral activity.

Particles of both solar wind and ionospheric origin can become trapped in certain regions of the magnetosphere close to the Earth, the *Van Allen belts*. Lower energy plasma evaporated from the ionosphere populates the *plasmasphere*, which lies under the outer Van Allen belt to a distance of about 4 Earth-radii. The plasma in the plasmasphere is relatively cool by magnetospheric standards, at a temperature of around 2000 K. Material in the plasmasphere co-rotates with the Earth below, unlike the more rapidly circulating particles in the Van Allen belts.

The Van Allen Belts

The motions of charged particles in a magnetic field are controlled by a number of influences. Important among these is the *Lorentz force*, which deflects charged particles entering a magnetic field at right angles to the field, and also at right angles to their previous direction of motion. The consequence of this "right-hand rule" is that particles spiral along magnetic field lines. Being of opposed charge, protons and electrons spiral in opposite senses around a field line.

In certain regions of the magnetosphere relatively close to the Earth, charged particles may become trapped, spiraling back and forth along closed magnetic field lines for long periods. The spirals described by trapped particles are quite open at larger distances from the Earth, where the magnetic field intensity is lower. In the stronger magnetic field closer to the Earth, however, the spirals tighten until a point is reached where the trajectory is exactly perpendicular to the field line. Particles are deflected back along the field line in the opposite direction from this "*mirror point.*"

The existence of trapping regions in the Earth's magnetosphere was predicted in theoretical studies by the Norwegian auroral physicist Carl Stormer (1874–1917) in the early twentieth century, and these were confirmed in measurements from the radiation detectors aboard Explorer 1 in 1958; they are now commonly known as the Van Allen belts.

There are two Van Allen belts. The inner, containing protons and electrons of both solar wind and ionospheric origin, lies around an average distance of 1.5 Earth-radii (9100 km) above the equator. The outer Van Allen belt contains mainly electrons from the solar wind, and has an equatorial distance of 4.5 Earth-radii (29,000 km). "Horns" of the outer belt dip sharply in toward the polar caps in either hemisphere. As a result of the offset between the Earth's geographic and magnetic axes, the inner belt reaches a minimum altitude of about 250 km above the Atlantic Ocean off the Brazilian Coast. This *South Atlantic Anomaly* occupies a region through which low-orbiting satellites frequently pass; problems can sometimes arise due to the energetic particles there.

Van Allen belt electrons typically have energies of a few MeV, while trapped protons may range from 10 MeV to 700 MeV: fortunately for those using satellites for communications, astronomical observations, or other purposes, protons in the higher range are rather rare. [A useful means of describing the energy of subatomic particles, the electronvolt (eV), is equivalent to the energy acquired when an electron is accelerated through a potential difference of one volt.]

Particles trapped in the Van Allen belts were once thought to play a primary role in governing auroral activity. It was believed that, when the right conditions prevailed, particles from the Van Allen belts could escape ("precipitate") into the high atmosphere, there giving rise to auroral activity. This model, however, lacks a satisfactory mechanism for accelerating auroral electrons to the observed energies. It has also been stressed that the total particle population in the Van Allen Belts would be insufficient to sustain major auroral activity for much longer than an hour or so. The most widely accepted modern theories suggest that the main magnetospheric reservoir of the particles that are subsequently accelerated and injected into the atmosphere to cause the aurora is in the plasma sheet.

In addition to their motion back and forth between the hemispheres, particles trapped in the Van Allen belts undergo a lateral drift, traveling around the Earth in longitude over timescales of minutes to hours: experimental releases and high-altitude nuclear test explosions have demonstrated that this drift rapidly distributes particles around the belts. Being of opposite charge, electrons and protons drift in opposite directions around the Earth: electrons drift eastwards, protons westwards.

The main influence of the Van Allen belt particle populations during auroral activity, according to more recent theories, is in decreasing the global magnetic field intensity, as they circulate faster in response to disturbed conditions. Ring currents in the Van Allen belts intensify markedly at the onset of geomagnetic storms, and contribute to disturbances of the global ground-level magnetic field at such times.

Particles in the Van Allen belts are trapped for long periods, but there is some degree of turnover, and they may be lost from the belts by several processes. Radio waves, from natural sources such as lightning discharges (producing "whistlers") or man-made sources, can transmit along closed terrestrial magnetic field lines looping out through near-Earth space. Resonances between these waves and trapped particles can reduce the particles' kinetic energies, and cause them to precipitate out of the trapping regions. More important, resonances may increase particle velocities along the field lines relative to their perpendicular velocities, resulting in the mirror point being moved closer to the Earth; if the mirror point is moved sufficiently close, particles will hit the atmosphere.

Collision between trapped particles and neutral hydrogen in the geocorona that permeates near-Earth space to a few Earth-radii is another mechanism by which material is lost from the Van Allen belts.

The particle concentration is also subject to change, largely in response to geomagnetic activity. Particle concentrations in the inner belt change only very slowly, by a factor of about three over the course of 12 months or so. The outer Van Allen belt particle population may change by a factor of ten in less than a day.

Plasma Movements and Currents in the *Magnetosphere; the* Auroral Ovals

Geographically, auroral activity is present more or less permanently at high latitudes, around the auroral ovals. These are rings of aurorae, asymmetrically displaced around either geomagnetic pole. Under quiet geomagnetic conditions, the auroral ovals remain relatively fixed in space, above the rotating Earth. Each auroral oval extends furthest toward the equator on the night-side, such that an observer at a given mid-latitude geographical location is carried closest to the oval around the time of local magnetic midnight. On the day-side, the auroral ovals appear pushed back towards the geomagnetic poles, reflecting the overall distortion of the magnetosphere resulting from its interaction with the solar wind.

Solar wind plasma entering the magnetosphere carries with it a frozen-in magnetic field derived from the IMF. An important process in driving the auroral mechanism is joining—*reconnection*—between the IMF and terrestrial field lines at the leading edge of the magnetopause upwind of the Earth. This merging process occurs with highest efficiency when the IMF is directed southwards relative to the ecliptic. A crude analogy is the joining together of two bar magnets by their opposed (N and S) poles. The reconnected field lines join the magnetosphere to the solar wind, resulting in transfer of energy across the magnetopause boundary.

Such joined field lines, and their entrained plasma, are dragged down the magnetotail by the solar wind. Plasma flows along the outside of the magnetotail lobes in a plasma mantle on the boundary of the magnetosheath. As it travels, the plasma is subject to the influence of electric currents in the magnetotail, experiencing *E-cross-B drift* perpendicular to the current and to the magnetic field. Both electrons and protons are deflected in the same sense, so the plasma flows together as an ensemble towards the equatorial neutral sheet.

Frozen-in magnetic fields will move with a plasma population provided there is no sudden change in local magnetic field direction or drop in its intensity. Where abrupt changes or weakening do occur, magnetic field lines are able to diffuse through the plasma and the frozen-in condition breaks down. Where such diffusion occurs, field lines of opposed magnetic polarity can come together and reconnect. Such a diffusion region exists in the magnetosphere, some 100 Earth-radii (637,000 km) downwind in the magnetotail under undisturbed conditions.

"Interplanetary" field lines, reconnected at the neutral line, sweep on down the magnetotail to rejoin the solar wind. Plasma jets are ejected both upstream and downstream from the neutral line. The jet directed upstream, back toward the Earth, helps to maintain the population in the plasma sheet.

The same cross-field currents that influence the movement of plasma in the magnetotail are also projected onto the auroral ovals, such that there is a general current (electron flow) descending onto the evening sector, across the oval, and back out into the magnetosphere on the dawn side. A lesser, secondary, current flows in the opposite direction. These currents flow along magnetic field lines, and the auroral ovals could, effectively, be considered as extensions of the magnetosphere reaching down into the ionosphere in the Earth's upper atmosphere.

Under undisturbed conditions, the auroral ovals are quite narrow, containing thin sheets of discrete aurora (perhaps of the order of 1 km wide), surrounded by a much broader region (500 km or so) of diffuse aurora. During substorms and geomagnetic storms, the ovals become broader and more complex in structure.

Quiet aurora at high latitudes is produced by a steady stream of particles precipitating out of the earthward extension of the plasma sheet around the polar cap. This quiet aurora may, from time to time, be enhanced by an increased particle flux into the atmosphere. Such increases occur at the time of *magnetic substorms*, fluctuations in the geomagnetic field resulting from frequent, small changes in the solar wind IMF in the near-Earth environment.

Much more dramatic are the *geomagnetic storms* that result from major variations in IMF intensity and direction. During these events, which usually follow violent activity in the inner solar atmosphere, the auroral ovals become greatly disturbed, broadening and expanding equatorwards, particularly on the night-side. Such events bring the aurora to the skies of middle latitudes. While substorms are of fairly short duration (a few hours), it may take several days for the magnetosphere to settle down following a major geomagnetic storm.

Magnetic substorms and geomagnetic storms represent periods during which the power of the magnetosphere–ionosphere circuit increases, resulting in brighter and more extensive auroral activity. Several commentators have used the analogy of a television set to describe this situation, with the upper atmosphere as the "screen" illuminated by electrons fired from the magnetotail "gun." The shifting pattern of auroral activity is the signature of large-scale interactions between the magnetosphere and the solar wind.

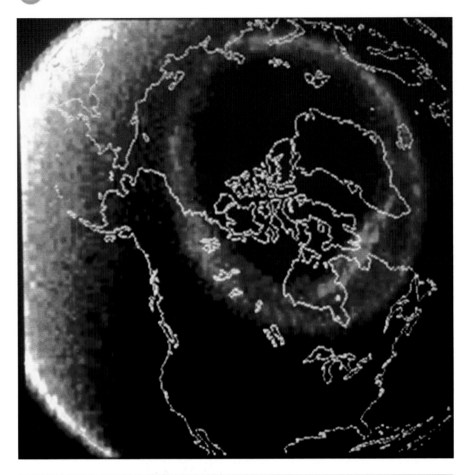

Figure 2.13. A Dynamics Explorer-1 view of the northern hemisphere auroral oval taken from high orbit. The outline of continental North America has been added to indicate the geographical extent of the aurora. A substorm is in progress, with a bright band of aurora over northeast Canada indicating the westwards-travelling surge. Image: University of Iowa.

Electric currents flow between the outer and inner edges of the auroral ovals, parallel to the Earth's surface. As a result of the sharp dip of the magnetic field at the high latitudes of the quiet-condition auroral ovals, these currents are directed *perpendicular* to the magnetic field, giving rise to an E-cross-B drift, with protons and electrons drifting from the day-side to the night-side.

At higher altitudes (300 km or so), protons and electrons drift at the same rate. Lower in the ionosphere (around 100 km altitude), the atmospheric density is higher, leading to a selective loss (via particle collisions) of protons. The result of this process is the establishment of electric currents flowing eastwards along the evening sector of the auroral oval, and westwards in the morning sector. These *auroral electrojets* in the ionosphere become enhanced at times of high geomagnetic activity. Where the

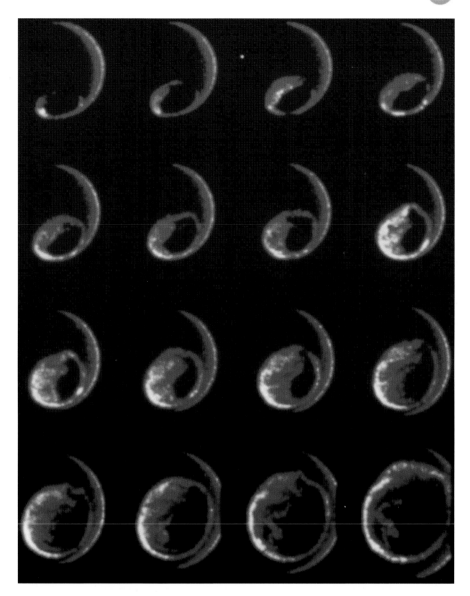

Figure 2.14. Progress of a substorm in the auroral oval as seen from above by the Dynamics Explorer-1 satellite. Images, taken in the ultraviolet, at intervals of several minutes show the initial brightening of the night-side of the oval, followed by the westwards-travelling surge, and expansion of activity to higher latitudes within the oval. Image: University of Iowa.

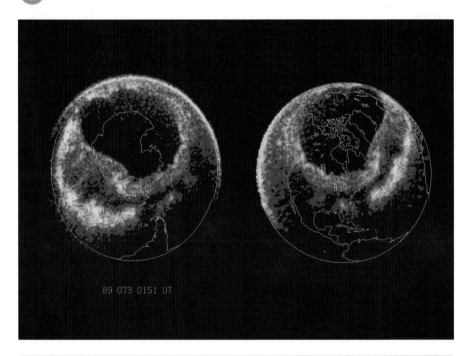

89 073 0151 UT

Figure 2.15. Dynamics Explorer image of the auroral oval during the Great Aurora of 13–14 March 1989, demonstrating the sheer extent of activity—for example, to Mediterranean latitudes and lower—during this extreme event. Image: University of Iowa.

eastward and westward electrojets meet, around the midnight point on the auroral oval, lies a region of turbulence, the *Harang discontinuity*.

The visible aurora results from collisions between energetic electrons and atmospheric particles. Electrons from the neutral sheet plasma carry insufficient energy to penetrate the atmosphere. As they travel along magnetic field lines into near-Earth space, these are subjected to the same mirroring experienced by electrons trapped in the Van Allen belts. Following the 1977 multi-spacecraft ISEE mission, however, a mechanism by which plasma sheet electrons could become accelerated, the *auroral potential structure*, became apparent.

The auroral potential structure is produced at times when the power of the magnetosphere–ionosphere circuit increases, as during magnetic substorms or geomagnetic storms. Thin sheets of positive and negative charge, aligned to magnetic field lines and lying close together, develop. Acceleration of electrons along these thin sheets offers an explanation for the aurora's appearance in narrow "draperies'" in the high atmosphere. Over distances of 10,000–20,000 km (1.5–3.0 Earth-radii), a potential drop of the order of kilovolts arises, down which electrons can be accelerated, gaining enough energy (several keV) to penetrate the atmosphere, and there produce the visible aurora.

The most direct route of entry for solar wind plasma is via the narrow *cusps* between field lines on the day-side of the magnetosphere at high latitudes. In these regions, the magnetic field lines are closely bundled together, and dip sharply toward the Earth's surface. Particles dribbling in by this route carry relatively little energy and consequently give rise only to diffuse, fairly pallid aurorae. The structure of the cusp regions has been clarified by satellite observations and measurements. Surrounding either cusp is a broader region, the *cleft*, which is also characterized by low-energy precipitation, but does possess some higher energy structural features.

Auroral Activity Types and Causes

Auroral Substorms

For much of the time, the aurora at high latitudes is relatively quiescent, in the form of narrow, faint arcs. Occasionally, activity may suddenly intensify and brighten, during *auroral substorms*. The occurrence of such events as global phenomena first became recognized following careful study of simultaneously taken photographs from ground stations, obtained during the International Geophysical Year in 1957–1958. The overall pattern of activity around the oval during substorms has been confirmed by satellite images from polar orbit.

The onset of substorm aurora is quite rapid, and first manifests as a brightening of the oval during what is termed the *growth* phase. Activity spreads both eastwards and westwards on the night-side of the auroral oval during the subsequent *expansive* phase, then begins to migrate polewards, resulting in the production of a wavy structure which moves westwards from the midnight sector into the dusk sector of the oval. This *westward traveling surge* moves at about a kilometer per second. Meanwhile, the morning sector of the oval breaks up into patchy rayed structures. The expansive phase may last for some 30 minutes. Once the polewards migration has reached its maximum extent (75–80° geomagnetic latitude), activity begins to subside, during the *recovery* phase.

Substorm aurorae are seen only at high latitudes, and result from relatively smaller geomagnetic disturbances than the major storm aurorae occasionally seen at mid-latitudes. There may be as many as four or five substorms, lasting between 1 and 3 hours, each day.

While the precise details remain to be completely understood, it is now widely accepted that the main driving force behind substorm activity is the condition of the solar wind in the Earth's vicinity. Substorm activity reflects the efficiency of coupling between the IMF and the terrestrial magnetic field. At times of most efficient coupling, typically when the IMF has a strong southerly direction, the aurora brightens and substorm aurora is seen.

The power generated by the auroral dynamo may be of the order of a million megawatts, and is associated with an electric potential of 100,000 volts, driving magnetospheric plasma into the upper atmosphere at high latitudes. Plasma is also ejected down the magnetotail and back into the solar wind in the form of *plasmoids*, magnetic "cage" structures, discovered as a result of studies using the ISEE-3 spacecraft in 1983.

Early models for substorm generation invoked spontaneous, explosive reconnection between opposed magnetic field lines, induced by the circulating currents in the lobes of the magnetotail. More recently, Edward W. Hones has proposed a model relating substorm aurora to periods of high southerly directed IMF. The onset of substorm activity is preceded by progressive stretching of magnetic field lines lying more than 7 Earth-radii (44,800 km) downstream of the Earth. This stretching is a consequence of more efficient reconnection between IMF field lines and field lines on the sunward leading edge of the magnetopause. In turn, the magnetotail acquires an excess of (solar wind-derived) energy.

The key event seems to be the spontaneous (and as yet unexplained) development of a new neutral line—the substorm *neutral line*—about 15 Earth-radii (96,000 km) downwind from the Earth. This substorm neutral line interferes with the cross-tail electric current in the neutral sheet, resulting in an abrupt collapse of the stretched magnetic field structure that has developed over the previous hour or so. This collapse is accompanied by the deposition into the atmosphere of large numbers of accelerated electrons from the magnetosphere, which gives rise to enhanced auroral activity.

Geomagnetic Storms

Auroral substorms at high latitudes arise when the magnetosphere is moderately disturbed, in response to relatively small changes in the solar wind IMF. The more major disturbances, which follow violent events associated with flares in the inner solar atmosphere, have more dramatic consequences. The geomagnetic storms that sometimes (but do not *always*) follow these events are much less frequent than the high latitude substorms, but can last for several days. Geomagnetic storms carry auroral activity to lower latitudes, and into the skies of the more populous regions of the world. These are the events witnessed by our forebears as "battles" or "dragons" in the sky, and of which the Great Aurorae of 1909, 1938, 1989, and 2003 are further, modern examples.

Such displays are usually commonest in the rising phase of a sunspot cycle toward maximum, when large, actively changing sunspot regions prone to flare production are often present on the solar disk and coronal mass ejections are frequent. Sunspot cycle 23, in its declining phase through the early 2000s, bucked this trend somewhat, producing its most vigorous and extensive storms late on; we shall return to this in a later chapter.

Coronal mass ejections, traveling out from the Sun at velocities of up to 1200 km/s, give rise to shock-waves in the solar wind. Compression of the magnetosphere by such shock waves intensifies the terrestrial magnetic field for a period of some tens of minutes, producing an effect detected using magnetometers at ground level as *Sudden Storm Commencement* (SSC). The onset of SSC can be an early indicator of the possibility of aurora visible at lower latitudes, but not all SSC events are necessarily followed by strong visible displays.

One consequence of the compression of the magnetosphere at the time of SSC is an intensification of the ring currents circulating in the Van Allen belts. The compression effect of impact between the magnetosphere and a shock wave propagating out through the solar wind also acts to intensify magnetic fields in the solar wind.

In the production of geomagnetic storms, the most important factor is the IMF orientation. Even very major CMEs will have little or no effect (other than production of SSC) if the IMF is directed northwards, and the efficiency of reconnection between solar wind and terrestrial magnetic field lines is low. Even on occasions when the shock is accompanied by a southerly directed IMF, it appears to be necessary for such conditions to prevail for many hours (Japanese auroral physicist Syun-Ichi Akasofu suggests a minimum of six hours) before a full-scale geomagnetic storm will be initiated.

The direction from which the shock-wave in the solar wind approaches the magnetosphere is also important. Flares associated with sunspot groups near the solar limb eject shocks more or less at right angles to the Sun–Earth line. The shock-wave in such instances will glance mostly *across* the leading edge of the magnetosphere, causing little compression, reducing the chance of initiating major auroral activity. If, however, the shock-wave comes more or less head-on toward Earth down the Sun–Earth line, the compression may be very major, and be followed by extensive auroral activity.

This was, indeed, the situation in the great March 1989 storm. Major flares associated with the progenitor spot-group occurred while the group was close to the limb: some flare-associated ejections were even detected while the spot-group was on the Sun's averted hemisphere, *behind* the limb! These had little effect on the Earth. A week or so later, when the group was near the Sun's central meridian as viewed from Earth, another flare occurred, whose associated shock-wave arrived, head-on, within 24–36 hours, producing the biggest auroral display seen for decades!

Under these conditions, the auroral ovals again brighten as the power of the auroral dynamo increases. A westward-traveling surge sets in, and the ovals expand toward the pole. This expansion is also accompanied, in geomagnetic storms, by an expansion *equatorwards* from the ovals, and the formation of multiple bright arcs on the nightside. The whole structure can become very extensive, bringing the aurora into the skies of mid-latitudes. "Gusty" conditions in the solar wind behind the flare-associated shock-wave may be manifested in the sometimes chaotic moment-to-moment changes in the nature of auroral activity during these storms.

Other Effects Associated with CMEs

Radio astronomers can detect "squalls" in the solar wind associated with coronal mass ejections as scintillation—twinkling—of quasars, which are extremely distant (and, therefore, compact) sources. Antony Hewish and colleagues carried out such observations from Cambridge between 1964 and 1981.

As we have seen, the Sun is a source of relatively low-energy cosmic rays in the form of protons accelerated during solar flare events. More energetic Galactic cosmic rays originate from beyond the Solar System, undergoing acceleration in violent events such as supernova explosions. Galactic cosmic rays are, chiefly, protons, but heavier ions are also represented in the population.

Cosmic rays arriving at Earth may be detected in a number of ways, including the use of Geiger counters, scintillation counters, cloud chambers, or photographic emulsions. Detectors have been flown on balloons and carried aboard satellites. Detection at ground level is dependent on the arrival of secondary products resulting from collisions between the primary cosmic ray and atmospheric particles—Geiger counters at Earth's surface detect electron *air showers* resulting from a secondary "cascade." Detectors at high-altitude observatories (Calgary, Canada, for example) can measure the flux of neutrons: increases in this associated with proton ejection during solar flares gives rise to Ground Level Enhancement (GLE) events.

The Galactic cosmic ray flux at Earth is generally reduced a sunspot maximum, when the strong solar wind their penetration to the inner Solar System. Interludes of still-further reduction are found in association with coronal mass ejection events. Passage of a CME through near-Earth space gives rise to a drop in Galactic cosmic ray flux known as a *Forbush decrease*; these events were first studied in detail by Scott E. Forbush and colleagues in 1942. The onset of a Forbush decrease coincides with arrival of the CME shock wave, seen in the geomagnetic record as Sudden Storm Commencement.

Coronal Hole Aurorae

The late-cycle persistent streams ejected into the solar wind from coronal holes give rise to geomagnetic activity in some respects intermediate between that of substorm and geomagnetic storm aurorae. As with the latter, a key element appears to be the velocity component. Coronal hole streams can typically reach velocities of 800 km/s. Compression effects as a coronal hole stream sweeps across the leading edge of the magnetosphere again lead to localized intensification of the IMF in the solar wind, in turn leading to enhanced auroral activity.

The passage of coronal hole streams brings about equatorwards expansion of the auroral ovals, though not to the extreme extent seen during geomagnetic storms. Also, the IMF in coronal hole streams is less prone to momentary changes or turbulence, such that auroral activity is also less violent. Characteristically, coronal hole aurorae penetrate to higher mid-latitudes (they are rarely, if ever, seen from, say, the latitudes of London or the southern United States), and consist of quiet arc or band structures, with little of the rapidly moving rayed activity seen during major geomagnetic storms. The intensity of coronal hole displays is also lower, with fainter, colorless forms predominating. Ron Livesey, for many years the distinguished Director of the British Astronomical Association Aurora Section, came to refer to these displays as "Scottish Aurora," in recognition of their more northerly visibility range—contrasting with the big storm events that may be visible to the latitudes of southern England.

Coronal holes may persist, relatively undiminished, for many months, with the resulting possibility of recurrent activity each solar rotation, as the stream sweeps across the Earth repeatedly. Such series of quiet aurorae at higher mid-latitudes, recurring every 27 days, are indeed common in the year or so before sunspot minimum, when the coronal holes are most common at lower heliographic latitudes. The recurrence of these relatively quiet aurorae was first described by the German astronomer Julius Bartels (1899–1964) in the 1930s, and until coronal holes were discovered several decades later, such activity was ascribed to (magnetic) "M regions" on the Sun.

As a result of the inclination of the Earth's orbit relative to the solar equator, a seasonal effect in the occurrence of coronal hole aurorae also operates. Earth is more closely aligned with coronal holes at more typical, higher, heliographic latitudes around the equinoxes. Coronal hole-associated aurorae are therefore most often seen around March and September.

Sector Boundary Crossings and Chromospheric Filament Disappearances

Short periods of enhanced geomagnetic activity can result from sector boundary crossings, which have already been discussed in relation to disconnection events in cometary ion tails. Rapid changes in the local IMF, similar to those that come on first encountering fast coronal hole streams, are seen. These effects are more noticeable at times of otherwise low solar activity, being masked at sunspot maximum by the turbulent conditions prevailing in the solar wind.

The disappearance of chromospheric filaments can also result in an enhancement (often relatively small) of the IMF passing the Earth, and be followed by increased geomagnetic and auroral activity. Again, these effects are most obvious in the declining years of the sunspot cycle. Disappearing filaments offer an explanation for some of the reasonable late-cycle auroral displays, which can extend to lower latitudes, in the apparent absence of candidate sunspot activity. It is also noteworthy that a disappearing filament appears to have been the most likely cause of one of the most major auroral storms of sunspot cycle 22, on the night of 8–9 November 1991.

Auroral Emissions

The principal atmospheric constituents involved in auroral emissions at altitudes typically of 100 km and upwards are nitrogen and oxygen, excited by energetic electrons from the magnetosphere's neutral, accelerated into the high atmosphere in response to fluctuations in solar activity, transmitted to the Earth via the solar wind. Oxygen and nitrogen are, of course, the main constituents of the atmosphere, but at auroral heights have a differing abundance ratio and nature from the 78% N_2 : 21% O_2 observed in the troposphere.

Above about 80 km altitude, ultraviolet radiation from the Sun dissociates molecular oxygen (O_2) into its single-atom, atomic form (often denoted as OI). The atomic oxygen concentration reaches a maximum at altitudes in the atmosphere of about 105 km. Relative proportions of gas species at 100 km are 76.5% N_2 : 20.5% O_2 : 3.0% OI. Atomic oxygen is predominant species, being approximately 100-fold more abundant than N_2, at 400 km altitude, while molecular oxygen is effectively absent above 130 km.

Ionized nitrogen, N_2^+, is also found in the high atmosphere, and is another product of the absorption of ultraviolet radiation from the Sun. In the auroral regions, collisions between N_2^+ ions and energetic electrons produce nitrogen atoms (NI).

Auroral light results from the excitation of atmospheric atoms and molecules during collisions with energetic particles, and subsequent re-emission of the imparted excess energy in the form of light. Spectroscopic studies of the aurora have revealed much about the nature of these emissions. The light of the visual aurora is produced almost entirely by collisions between electrons and atmospheric oxygen and nitrogen. The depths to which incoming accelerated particles of different energies can penetrate the atmosphere are indicated in Table 2.3.

Table 2.3. Penetration of Energetic Particles into Earth's Atmosphere

Particle	Energy (keV)	Penetration (km altitude)
Electron	1	150–200
	10	100
	30	90
Proton	30,000	50
	500,000	Ground level during severe polar cap event

For a given chemical species, there is a series of permitted energy levels (predictable from the laws of quantum mechanics) surrounding the positively charged nucleus in which electrons may orbit. Under normal circumstances, electrons sequentially fill the lowest energy levels (closest to the nucleus) first. Electrons may, at the expense of appropriate quanta of energy, be transferred between levels—that is, undergo *transition*. Transitions to higher-energy (outer) orbital levels require the addition of energy from outside: in the case of atmospheric oxygen and nitrogen during auroral conditions, this is imparted by incoming electrons or protons accelerated in the magnetosphere, or from other atmospheric particles which have themselves previously undergone *excitation*.

Sufficient energy may be delivered to produce electron transitions to levels that cannot normally be occupied, since the presence of electrons therein could violate the quantum mechanical rules which demand that lower levels must be filled first. These transitions are therefore described as "forbidden." Emissions resulting from forbidden transitions are seen only in conditions of rarefied particle density, such as obtain at high altitudes in Earth's atmosphere.

After a given time interval, the excitation is lost as the electron drops back to a lower orbit, and the atmospheric particle returns to its preferred minimal-energy *ground state*. The excess energy is re-emitted as a photon of light whose wavelength is precisely governed by the same quantum rules that dictate the available energy levels surrounding a given nucleus. Figures 2.16 and 2.17 present some of the predicted spectral emissions associated with transitions between excited and ground states in atmospheric oxygen and nitrogen during auroral activity.

The presence of Doppler-shifted Balmer lines in auroral spectra indicates the involvement of protons (*hydrogen* nuclei) in producing some of the excitation which results in auroral emissions. Hydrogen emissions are commonly detected at altitudes around 120 km.

Visually, the aurora may often appear white or colorless when faint. Bright displays can show marked color, however, notably greens and reds. The dominant feature of the visible auroral spectrum is the forbidden green emission at 557.7 nm, resulting from the excitation of atomic oxygen (OI). This characteristic "auroral green line" may be detected using narrow-passband interference filters, allowing the observation of activity in moonlit or light-polluted skies, or in cloudy conditions.

Green atomic oxygen emission is dominant in the lower parts of auroral displays, around 100 km altitude. Higher in the atmosphere, collision of lower-energy electrons with atomic oxygen gives rise to red emissions at 630.0 nm and 636.4 nm wavelengths.

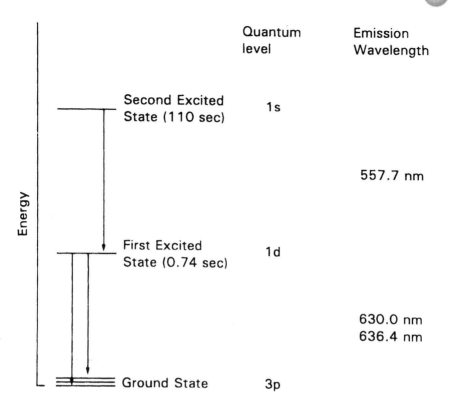

Figure 2.16. Energy level diagram for atomic oxygen, showing the electron transitions commonly found in aurorae and the light emission wavelengths produced when excited electrons return to lower energy states. Lifetimes in the excited states are also indicated.

Atomic oxygen has two principal excited states that give rise to visual auroral emission. The 557.7 nm emission results from the second, higher excited state, which requires a greater energy input. The lifetime of OI in the second excited state is short (0.74 second), and electrons in this energy level rapidly decay to the lower level of the first excited state, yielding photons at 557.7 nm wavelength in the process (Fig. 2.17). Red oxygen emissions result from the descent of electrons from the first excited state to ground state.

The lifetime of the first excited state of OI is longer (110 seconds) than that in the second excited state, leading to the observed variations in the predominant emission with height. In the lower, denser parts of the auroral layer, the slow red oxygen emissions are *quenched*: the time required for an excited electron to return to ground state is greater than the average interval between collisions of oxygen with other atmospheric particles. The excitation energy is therefore lost through collisions before it can be re-emitted as light. Conversely, the green auroral emission occurs rapidly, before most excited oxygen atoms can give up their excess energy through collision.

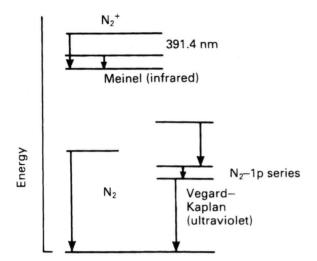

Figure 2.17.
Energy level diagram for molecular nitrogen (N_2), and ionised molecular nitrogen (N_2^+). The 391.4 nm emission is enhanced by resonance scattering in sunlit aurorae, and appears bluish-purple to the eye. Sunlit aurorae are often seen at high latitudes during late spring or early autumn.

The production of two red emission lines following the decay of electrons from the first excited state in atomic oxygen results from the availability of two, almost equivalent, energy levels in the normal outer quantum shell around the nucleus, either of which may require filling to restore the ground state configuration.

At altitudes of 1000 km, where the uppermost parts of auroral rays may reach above the Earth's shadow at certain times of year, molecular *nitrogen* is substantially ionized to N_2^+ by the action of solar ultraviolet: approximately 75% is in this form at the top of the auroral layer, compared with 20% of atmospheric nitrogen at 100 km altitude. Excitation of N_2^+ by incoming accelerated electrons produces blue–purple emissions at 391.4 nm and 427.8 nm wavelengths. N_2^+ emissions are also present in auroral features at altitudes around 100 km. In sunlit aurora, however, the blue–purple emissions are enhanced by absorption and re-emission of these wavelengths from solar radiation, a process of *resonance* that results in the 391.4 nm and 427.8 nm lines being stronger than they appear when produced by the collisional mechanism alone.

Red auroral emissions seen on the undersides of arcs or bands in the most vigorous aurorae are produced by very energetic (30 keV) electrons penetrating to lower levels—of the order of 90 km altitude—and exciting molecular nitrogen to emit a group of four spectral lines between 661.1 and 686.1 nm wavelength.

All these emissions can, at times, be so weak that the visual observer is unaware of their occurrence. Time-exposure photography using fast films and narrow-passband interference filters matched to auroral emissions can be used to determine the extent of low-energy auroral emission on a given night.

Table 2.4 highlights some of the important visually observed auroral emissions.

Balloon, rocket, and satellite measurements have extended professional auroral studies into the ultraviolet and infrared regions of the spectrum, in which such as the N_2 Vegard-Kaplan and N_2^+ Meinel emission are produced, respectively. Emissions from atomic nitrogen (NI) also appear in these regions. Detectors aboard Dynamics

Table 2.4. Some Visual Auroral Emissions

Wavelength (nm)	Emitting Species	Typical Altitude (km)	Visual Color
391.4*	N_2^+	1000	Violet–purple
427.8	N_2^+	1000	Violet–purple
557.7	OI	90–150	Green
630.0	OI	>150	Red
636.4	OI	>150	Red
656.3	H-alpha	120	Red
661.1	N_2	65-90	Red
669.6	N_2	65-90	Red
676.8	N_2	65-90	Red
686.1	N_2	65-90	Red

* Also at 50–70 km during severe polar cap absorption events

Explorer-1 recorded an ultraviolet emission of atomic oxygen at 130.4 nm in order to produce images of the auroral ovals taken from above the poles.

Other Solar Effects on the Atmosphere

The Ionosphere

An important effect of the action of sunlight on the Earth's upper atmosphere is the break-up of oxygen and nitrogen molecules—photodissociation—and subsequent *ionization*, of the free atoms. The chief components of sunlight responsible for this ionization are short wavelength emissions in the X-ray and ultraviolet regions of the spectrum. This ionization produces atmospheric populations of positively charged ions and negatively charged electrons.

Ionization appears at a number of atmospheric levels, producing layers that may be identified by their interactions with radio waves. Collectively, the high atmosphere layers of ionization at heights above 60 km or so are referred to as the *ionosphere*. The first ionospheric layer to be recognized (in the 1920s) was the *Heaviside layer*, now more commonly known as the *E-layer*. The E-layer is used by radio operators as a reflective surface from which signals can be "bounced" to distant stations lying over the horizon, and beyond range of horizontally transmitted ground waves. Short-wave radio reflection from the ionospheric E-layer shows a diurnal variation, being enhanced during the hours of darkness. This effect is a consequence of the increased absorption of radio waves during daytime, when sunlight produces higher electron densities.

The E-layer lies at an altitude of about 110 km (similar to the base atmospheric level of the visible aurora), and is present over the whole Earth. The higher F-layer shows a partial split into two regions, a lower F_1 region (at about 160 km altitude), and upper F_2 region (300 km). Diurnal effects again operate. The F_2 layer disappears at night, and may also disappear during daytime on occasion.

The F-layer, like the E-layer, reflects radio waves. The lowest region of the ionosphere, the D-region between 65 and 80 km altitude, however, principally *absorbs* radio waves. D-region absorptions occur in response to solar activity, and are frequently noted at the time of sunspot maximum. Both ultraviolet and X-rays from the Sun produce ionization in the upper D-region. The X-ray flux is more variable, and is also the major source of variation in D-region ionization. Increased ionization of the D-layer, produced by the enhanced emission of X-rays during solar flares associated with large, active sunspot groups, gives rise to Sudden Ionospheric Disturbance (SID) events. SIDs affect only the day-side D-region, and bring about abrupt increases in radio absorption. These events may last a matter of minutes, up to an hour or so. Ionization produced during SIDs is fairly rapidly lost as electrons recombine with positive ions in the relatively dense D region of the atmosphere.

Atmospheric absorption prevents penetration of X-rays below about 60 km, while the stratospheric ozone layer—fortunately for living organisms at the Earth's surface—significantly cuts off ultraviolet penetration below around 50 km altitude. Ionization in the lower parts of the D-region is caused by the penetration of energetic cosmic rays. An apparent paradox, resolved on development of models of the solar wind and heliosphere (below), was the observation that lower-level D-region ionization reaches a minimum at sunspot maximum, when upper D-region ionization peaks. It is now known that the flux of high-energy Galactic cosmic rays reaching the inner Solar System is reduced by the increased heliospheric magnetic field intensity around sunspot maximum.

Polar cap absorptions, again in the D-region, are an important cause of short-wave radio blackouts at high latitudes. These absorptions result from an increase in ionization of the middle atmosphere by highly penetrative protons ejected during solar flare events, and are closely associated with auroral phenomena.

Auroral activity can have a number of effects on the ionosphere, leading to disruption of short-wave radio communication. There are times, however, when auroral conditions in the ionosphere are *beneficial* to radio communication, allowing longer than normal distance contacts to be made (Chapter 3).

Photodissociation of atmospheric components by sunlight is, obviously, a daytime phenomenon. The mechanisms by which ionization is lost are important in shaping the ionosphere. In the absence of continued solar excitation, many of these loss processes operate at night, involving recombination between electrons and positive ions. Some of these recombination processes release the diffuse, weak background light of the airglow (Chapter 8).

The Heliosphere

We began this chapter by looking at our Sun's variability, and the suggestion that this might influence not just auroral and geomagnetic activity, but also terrestrial climate—could Space Weather and meteorology really have more in common than

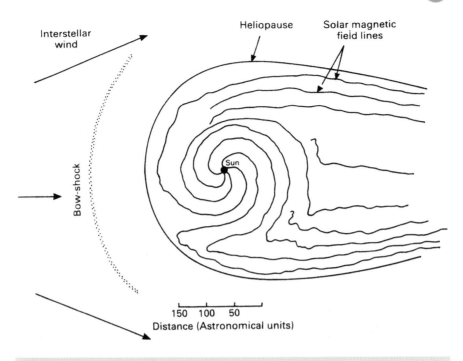

Figure 2.18. The region of space in which the solar magnetic field, carried outwards with the solar wind, is dominant is the heliosphere. From data returned by interplanetary probes, notably the Voyagers, the outer boundary—the heliopause—is found to lie close to 100 astronomical units from the Sun. Elements of the heliosphere's gross structure can also be inferred from spacecraft measurements, and from observations of comets. The solar wind magnetic field initially emerges more or less radially (four magnetic sectors are shown), becoming wound into a spiral with distance from the Sun. The postulated interstellar wind draws the heliosphere out into a comet-shaped structure which may be compared with the Earth's magnetosphere (Fig. 2.12).

may at first be apparent? To address the question, it is necessary to first look at the Sun's extended realm of influence, the *heliosphere*.

The heliosphere can in some respects be viewed as the Solar System-scale equivalent of a planetary magnetosphere. In other words, it is the volume of space in which the Sun's magnetic field has a dominant influence on particle motions, embedded within the still more extended Galactic environment of our interstellar neighborhood, in a spiral arm in the outer parts of the Milky Way.

In a very real sense, Earth—in common with the other major planets of the Solar System—orbits within the tenuous extended outer atmosphere of the Sun. The full extent of the region of space in which the Sun's influence is dominant has only really become apparent in recent years, with the advent of deep space probes—launched primarily to explore the outer planets. Valuable data have been returned by the Pioneer 10 and 11, and Voyager 1 and 2 spacecraft launched in the 1970s.

To establish the heliosphere's extent, scientists seek detection of the *termination shock*, where the solar wind rapidly becomes decelerated as it comes up against the interstellar medium. The termination shock lies inside the *heliopause*, the outer boundary of the heliosphere. Early models placed the boundary perhaps 10 AU from the Sun, around the same distance as Saturn's orbit. Following encounter with the ringed planet, however, the Voyagers continued to detect the solar wind, and right through the 1990s—long after they had crossed the orbit of Pluto, 40 AU from the Sun—there was no evidence that the termination shock was nearby.

First indications that the Voyager probes might be approaching the outer limits of the heliosphere came from measurements obtained between July 1992 and early 1993, when low frequency (1.8–3.5 kHz) radio waves were detected, consistent with interactions between material from coronal mass ejections and the interstellar medium at the heliopause. The time-delay between CMEs and production of radio noise suggested a heliopause distance of 110–160 AU from the Sun. Further signs that Voyager 1, the greatest-traveled space probe, was approaching the heliopause came in 2002, at a distance of 85 AU from the Sun. Then, late in 2004, Voyager 1's instruments began to detect tangible evidence, picking up the signature of *anomalous cosmic rays*—particles of interstellar origin, accelerated at the termination shock—at a distance from the Sun of 94 AU.

Much as Earth's magnetosphere "breathes" in and out in response to changes in solar wind velocity, it seems that the distance between the termination shock and the Sun responds to varying solar activity. Scientists believe that Voyager 1's December 2004 passage into the termination shock was brought about partly by the latter's inward movement in the declining phase of the sunspot cycle as the net solar wind intensity decreased. From late 2004 onwards, Voyager 1 has been considered to lie in the *heliosheath* beyond the termination shock, but still inside the heliopause. Once it crosses the heliopause, in perhaps 10–20 years, it will become mankind's first truly interstellar probe.

The heliosphere is, then, a very extended structure, with its boundary at least 100 AU from the Sun. Just as Earth's magnetosphere is shaped by the solar wind flowing past it, the heliosphere is thought to be sculpted by a Galactic interstellar wind, to form a teardrop structure with an upwind bow shock. The Solar System as a whole has a proper motion of about 20 km/s towards the direction of the constellation Hercules, and it is in this direction that the heliospheric bow shock might be expected to lie. The Voyagers and Pioneer 10 are leaving the Solar System toward the bow shock (Voyager 2 is expected to reach the termination shock in 2009 or 2010), while Pioneer 11 is heading down the tail of the heliosphere.

Effects on Terrestrial Climate

The flux of Galactic cosmic rays reaching Earth is modulated by the solar wind intensity. At times of high solar activity, when the overall solar wind velocity is greater, fewer Galactic cosmic rays can penetrate the heliosphere to reach the inner Solar System.

Galactic, high-energy cosmic rays colliding with atmospheric nitrogen are, as we have seen, the source of ^{14}C, and there is a clear anti-correlation between ^{14}C levels and solar activity. High-energy cosmic rays can also act as condensation nuclei for cloud formation, and it is through this role that terrestrial climate may be affected by the state of solar activity. A greater flux of Galactic cosmic rays means more cloud formation. Under the weakened solar wind conditions prevailing during the sunspot-sparse Maunder Minimum, for example, it is possible that a prolonged interval of increased Galactic cosmic ray flux could in time bring about a significant increase in global cloud cover. In turn, this would cause a decrease in the amount of solar radiation received at ground level—more is reflected back into space from the clouds. This mechanism, operating over decades, might account for the harsher conditions experienced during the Maunder Minimum and other episodes where low sunspot numbers have apparently coincided with periods of global cooling.

Conversely, during periods of high sunspot activity (and, therefore, diminished Galactic cosmic ray flux) such as the 12th century Medieval Maximum, global temperatures appear to have been elevated. Over the longer term, at least, it appears that solar activity can indeed have an influence on terrestrial conditions beyond the geomagnetic and auroral effects discussed above.

CHAPTER THREE

Auroral Forecasting

The growth, in recent decades, of "astronomical tourism" has led to some extent to a greater expectation—at least among those who can afford to travel!—that astronomical events and phenomena should be predictable, allowing trips and observations to be planned in advance. Total solar eclipses are the prime example, but the success of Asher and McNaught's Leonid storm forecasts (Chapter 1) also brought meteors into the same category. Improved understanding of how they are caused means that major, extensive auroral events can also be forecast with better accuracy, enabling would-be observers to get themselves out from under urban light pollution to locations with clear dark horizons for optimal viewing.

Of course, there is always the option of taking a holiday at a favorably placed high latitude location where the aurora is a nightly occurrence. Popular destinations are Alaska, or Tromso in Norway—but remember to go in winter when the sky is *dark*, avoiding the perpetual daylight of the Arctic summer! Brief "Northern Lights Flights" to see the aurora—flying from northwest Europe up towards the latitudes of the aurora oval and back in a single trip—have become popular, too. Passengers making night-time flights on transatlantic routes at high latitudes also often have the chance of the aurora as in-flight entertainment: obtaining a starboard side window on the flight from Europe to North America, or port on the trip from America to Europe, can be rewarding!

For most observers, most of the time, however, it is usually a case of waiting for the aurora to come to us. Even at quite low latitudes, the aurora will oblige several times during each sunspot cycle, and in this chapter we shall examine the circumstances under which this can happen, allowing the observer to be on alert at the right times.

Aurorae and the Sunspot Cycle

Interludes such as the Maunder Minimum discussed in the previous chapter apart, sunspot activity seems to be reasonably predictable, at least in terms of the timing, if not the intensity, of the periods of maximum and minimum activity. Sunspot cycles are numbered in sequence, starting from 1749 when systematic, daily solar observations began at the Zurich Observatory. Cycle 23 peaked in 2000–2001, and cycle 24 is expected to do so around 2011–2012.

For a very long time, it has been known that aurorae are more frequently seen at lower latitudes during periods of high sunspot activity. One might think, therefore, that in the long term the best time to be on alert for aurorae is at the maximum of the roughly 11-year sunspot cycle. The relationship between sunspots and aurorae is, however, rather more subtle. Figure 3.1 gives a comparison between six-monthly averages of sunspot group active area (AA) numbers, and the frequency with which aurorae were reported to the British Astronomical Association Aurora Section by observers in the British Isles south of the Orkney islands: for the aurora to be visible to those in mainland Britain—even Scotland—the auroral oval has to have undergone a reasonable degree of equatorwards expansion in response to disturbed geomagnetic conditions As can be seen for sunspot cycles 21 and 22—peaking in 1980 and 1990 respectively—lower-latitude aurorae are commonest perhaps a year or so before sunspot maximum occurs.

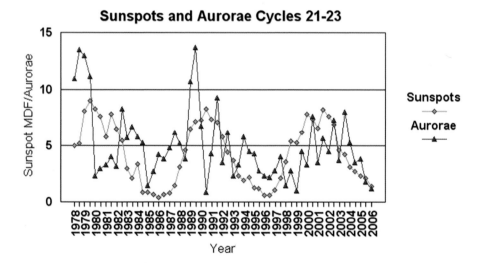

Figure 3.1. Six-monthly averages of sunspot MDF and occurrence of aurorae visible from the mainland British Isles in sunspot cycles 21, 22 and 23. In cycles 21 and 22 (sunspots peaking in 1980 and 1990 respectively), a clear primary peak in auroral activity about a year ahead of sunspot maximum is seen, with a secondary, less-intense auroral peak following the sunspots by 12–18 months. In cycle 23, however, the primary auroral peak seems absent, though there is a sustained 'tail' of strong auroral activity after sunspot maximum.

At sunspot maximum itself, there is a dip, then a second, somewhat less intense auroral peak follows about 18 months later.

This pattern has led to the suggestion that the solar wind/interplanetary magnetic field is much more disturbed in the run-up toward sunspot maximum. A magnetically "slack" interlude occurs at sunspot maximum, during which the solar magnetic field turns over. Establishment and strengthening of the new field orientation later in the cycle leads to the secondary auroral maximum.

The situation may not be so clear-cut, though, as shown by cycle 23. Cycle 23 beginning from the sunspot minimum of 1996/7 was notably slow to get started in terms of low-latitude auroral activity—indeed, the cycle did not produce a major event until the big 6–7 April 2000 storm. Sunspot numbers hit an extended, double peak in 2000 and 2002, but there is little suggestion of the early-cycle, higher auroral maximum found in the preceding cycles. Instead, cycle 23's main interval of active, low-latitude aurorae came in its declining phase. Even in the general, longer term, aurorae can be awkward to forecast!

The very major auroral storms—events like those of 13–14 March 1989, 6–7 April 2000, or 30–31 October 2003, which go into the record books as "Great Aurorae"—do, as a whole, cluster around the sunspot maximum, as shown in Figure 3.2. Plotted here against monthly sunspot AA values are those aurorae sufficiently intense to have

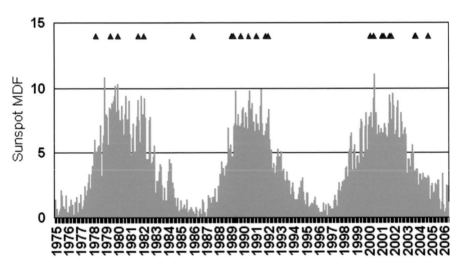

Figure 3.2. Monthly sunspot MDF figures for cycles 21 to 23, showing the marked fluctuations that can occur; multiple maxima are evident, together with interludes of lower activity—the rise an fall of the sunspot cycle is far from smooth. Plotted as the aurorae have been plotted as triangles across the top of the figure are major auroral storms (events visible to southern Europe, or the southern United States, for example). These generally cluster around the sunspot maxima, but note the exceptional case of the February 1986 storm, occurring almost at sunspot minimum. Use of auroral records to trace pre-telescopic solar activity can only be carried out with caution!

reached down to the latitudes of, say, the English Channel or the far southern United States. These are the auroral events that make headline news and are witnessed not just by amateur astronomers but also the general public. In every case, these aurorae have their origins in the most violent solar flares and coronal mass ejections, as the solar magnetic field releases immense stresses.

Astronomers attempting to reconstruct pre-telescopic solar activity have used major aurorae—recorded in Chinese or Korean annals, and in European monastic chronicles, for example—as tracers of sunspot maxima In most cases, this is probably a reasonable assumption. However, the major aurora of 8–9 February 1986, visible down to the tropics, almost at the sunspot minimum between cycles 21 and 22, serves as a reminder that caution must be exercised in interpreting the results.

Keeping an Eye on the Possibilities for Auroral Activity—The DIY Approach!

As we will see shortly, sources of data on interplanetary magnetic conditions, solar and geomagnetic activity, and, indeed, fully-fledged auroral forecasts, are numerous and increasingly accessible via the Internet. Amateur astronomers are, however, notably conservative in their approach, and many traditionalists like to carry out their own solar monitoring, for example. This is a perfectly valid option, especially if access to Internet forecasts is limited.

Solar Observing

The easiest, and safest, means of following day-to-day solar activity is by the widely used method of projection. This is one are a of astronomical observation where light pollution is not a problem, and in the case of the Sun the target object is so large and intensely bright that even a small refractor (this does not even need to be a high-class, top-of-the-range model!) in the 60–70 mm aperture bracket is quite sufficient to give worthwhile views.

Projection of the solar disk onto a white screen, shaded from extraneous light, will reveal dark sunspots and—near the solar limb—bright faculae. With a 70 mm refractor, it is possible to obtain a satisfactory 150 mm diameter projected image, big enough to reveal even quite small dark pores on the photosphere. On days when your local atmospheric conditions are steady—good astronomical *seeing*—the mottled pattern of photospheric granulation may even be apparent.

Traditionally, amateur astronomers have recorded daily solar activity by making drawings of the projected disk, showing the relative positions and numbers of sunspots. By this means, the growth and decay of spotgroups, and their progress across the disk as the Sun rotates, can be followed. Outlines of faculae can also be recorded.

The smallest sunspots are little more than simple pores, which appear as dark dots against the surrounding bright photosphere. Larger spots will show some

Figure 3.3. Projection through a small telescope is the simplest and safest way to observe sunspot activity.

structure—a dark central *umbra*, surrounded by lighter *penumbra*. In more complex spotgroups, there may be several umbrae and extensive surrounding penumbral regions.

While no two spotgroups are ever identical, a broad scheme is used to describe morphological types. This Zurich Classification is outlined in Table 3.1. Complex class D or E spots, in which umbrae appear to develop over the course of several days carry the potential to be sites of solar flare activity, and as these, particularly, cross the visible hemisphere of the Sun there is a possibility—but no guarantee—that lower-latitude aurorae may occur.

As a basic index of sunspot activity, observers estimate the numbers of active areas (AAs). An active area may be a single small spot, or a large, complex grouping of several spots covering a substantial part of the photosphere. To be counted as separate active areas, spots or spotgroups have to be separated by 10° on the solar circumference.

At the end of the month, the observer tallies up the total number of AAs seen, and divides by the number of days on which observations were taken to arrive at a Mean Daily Frequency (MDF). Table 3.2 presents an example from my observing log, dating back to April 1980—close to the maximum of sunspot cycle 21. Note that the fickle Scottish weather, under which these observations were made, results in less-than-total coverage; pooled results from geographically widely spread observers, collected together by bodies such as the BAA Solar Section or AAVSO will always allow the most

Table 3.1. Zurich Sunspot Classification

A – Pores, lacking penumbra. No magnetic bi-polarity.
B – Two or more spots, showing magnetic bi-polarity, but no penumbra.
C – Two or more spots, showing magnetic bi-polarity, and penumbra around either the leader or the follower.
D – Two or more spots with magnetic bi-polarity, and penumbra associated with both leader and follower. They cover up to 10° in longitude on the Sun.
E – As D, but covering 10–15° in longitude.
F – As D or E, but covering more than 15° in longitude.
G – Decayed remnant of D, E, or F.
H – Decayed remnant of C, D, E, or F; a single spot-group and penumbra. Larger than 2.5°.
J – As H, but smaller than 2.5° in diameter.

complete coverage, and individual amateur astronomers are strongly encouraged to submit their monthly observations to these organizations.

Month-by-month, year-by-year plots of MDF show the rise and fall of the sunspot cycle. The plot is usually far from smooth, with both relatively active and relatively quiet periods superimposed on the overall trend. Active periods, when sunspots are abundant—and especially if the emerging spotgroups are complex—are, naturally, of greatest interest to the would-be low-latitude aurora-watcher.

Another useful index of sunspot activity, favored by professional solar astronomers, is the Relative Sunspot Number , R—sometimes known as the Zurich, or Wolf Number

Table 3.2. Sunspots in April 1980

1980 April	AAs North Solar Hemisphere	AAs South Solar Hemisphere	Total AA Count
3	2	1	3
4	3	2	5
5	5	2	7
6	5	3	8
7	5	3	8
8	4	3	7
9	4	3	7
12	3	4	7
13	2	5	7
16	2	4	6
18	2	3	5
19	2	3	5
20	2	6	8
23	1	3	4
24	1	5	6
26	1	5	6
28	1	5	6

Days observed = 17
MDF North = 45/17 = 2.64
MDF South = 60/17 = 3.52
Total MDF = 105/17 = 6.18

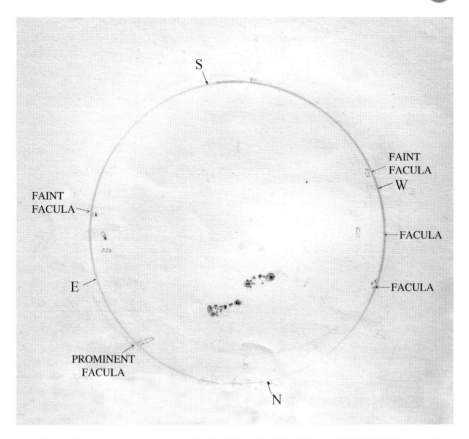

Figure 3.4. Simple drawing the Author of the solar disk, as projected through a small (40 mm aperture!) refractor on 7 April 1980, at a time of high sunspot activity. Two large, complex active areas were near the central meridian in the southern solar hemisphere, and several areas of facula were visible close to the Sun's limb. An AA count of 8 (5 northern hemisphere, 3 southern) was made from this observation.

after Rudolf Wolf who introduced it in 1848. R is given by a simple formula:

$$R = k(10g + f)$$

where k is the efficiency of the telescope/observer combination (usually taken to be 1, unless an extremely small telescope is used!), g is the number of spotgroups, and f is the total number of individual spots in all the visible groups. At times of high sunspot activity, as in March–April 2001, R may be as high as 330; close to sunspot minimum in 2006, more typical values were of the order of 30.

Active areas are most likely to influence solar wind conditions in near-Earth space when aimed toward us or—as a consequence of the "Parker spiral" structure of the Interplanetary Magnetic Field—somewhat to the west of the Sun's apparent central meridian (the line connecting north and south solar poles). In such positions, AAs are sometimes referred to as "geoeffective." As a crude rule of thumb, it takes about 48 hours for a disturbance caused in the inner solar atmosphere above an active area to reach Earth, so the presence close to the central meridian of a large, rapidly changing spotgroup *might* be considered as increasing the possibility of low-latitude aurorae a couple of days later.

An important caveat is that not every large, apparently active spotgroup will necessarily trigger major aurorae. Any number of large, potentially geoeffective AAs have come and gone in the past without troubling Earth's magnetosphere, and the presence of large sunspots is not in itself a guarantee of enhanced auroral activity. As discussed in Chapter 2, certain conditions have to prevail in near-Earth space for the auroral ovals to be pushed equatorwards, and not every disturbance propagating through the solar wind has the right characteristics.

Projection of the "white light" solar disk is certainly the most accessible means of following the Sun's state of activity for most amateur observers. Professional solar observatories, of course, have long monitored at specific wavelengths of the electromagnetic spectrum: as described in Chapter 2, observations in the light of hydrogen-alpha allow detection of solar flares, prominences and other phenomena of the chromosphere.

Until comparatively recently, such observations were somewhat beyond the reach of most amateurs. The 1990s, however, saw the appearance on the commercial market of dedicated hydrogen-alpha telescopes designed for amateur use and budgets. These allow the patient observer to watch the development of solar active regions in this wavelength, and even to witness, in real time, the occurrence of flares. The obvious point is that if a flare is seen in a spotgroup near the Sun's central meridian, then there is a chance that any accompanying coronal mass ejection may arrive in near-Earth space within the following 24–48 hours, and the possibilities for enhanced auroral activity are therefore good. As with much technology, mass production has brought about a gradual reduction in cost, and many amateur observers have taken to regularly using relatively affordable instruments like Coronado's PST (Personal Solar Telescope).

Recent Past Activity

Recent reports of auroral sightings can be of some value in attempting to anticipate aurorae in the immediate future. Where available—the Internet, and rapid-publication magazines such as *The Astronomer* are good sources—such reports may be plotted on a Bartels chart. Julius Bartels used these to plot out auroral and geomagnetic activity in 27-day strips, each corresponding to a single rotation of the Sun as observed from Earth, starting from 8 February 1832. Bartels rotation 2360 commenced on 28 April 2006, for example. Note that the Bartels rotations do not correspond with the Carrington solar rotation series that commenced on 9 November 1853.

The simplest form of Bartels diagram plots auroral events, coded for intensity, by date. Recurrent events will line up in adjacent strips. Particularly in the later parts of the sunspot cycle, as coronal holes become more common, series of quiescent aurorae may become apparent, spanning several solar rotations due to the persistence of the causative high-speed stream in the solar wind. Events associated with solar flare/CME activity are less likely to recur, as the active regions in which they arise are relatively short-lived.

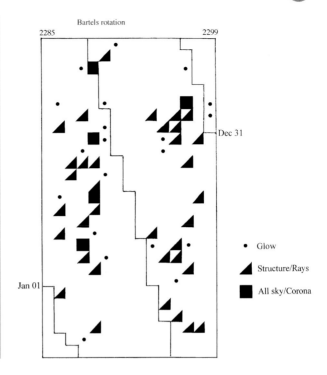

Figure 3.5.
Bartels diagram for 2001, showing UK auroral sightings during this very active year, close to sunspot maximum. Several major, all-sky or coronal storms were seen. Such events tend to be non-recurring, and so activity appears in only a single solar/Bartels rotation.

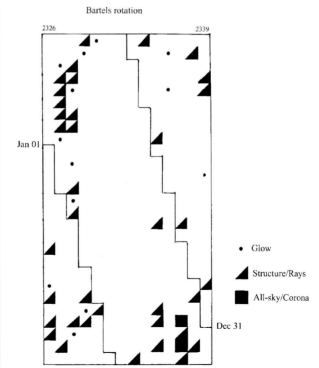

Figure 3.6.
Bartels diagram for 2004, in the declining phase of sunspot cycle 23. By this time, major storms were rare, but series of coronal hole-induced aurorae, with relatively low-level activity sustained over several successive nights, and recurring 27 days later became evident early in the year.

More complicated Bartels charts may be constructed to include, for instance, geomagnetic or radio propagation data.

Early Warnings of Ongoing Auroral Activity

Magnetic Effects

Disturbances of the terrestrial magnetic field in association with auroral storms were noted as early as the eighteenth century, using simple equipment. The technically adept twenty-first century amateur astronomer can obtain direct measurements and early warning of the onset of a geomagnetic storm using home-built *magnetometer* equipment.

The magnetic field at ground level may be described by three principal characteristics. These are D, the *angular deviation* (corresponding to the east–west swing of the compass needle relative to the magnetic pole, which it seeks under undisturbed conditions), H, the *horizontal field strength*, and Z, the *vertical field strength*. These characteristics all vary under disturbed geomagnetic conditions which may be accompanied by auroral activity. Field strengths are expressed in nanoTeslas (nT), equivalent to the geophysicists' unit, the gamma: 1 nT = 0.00001 gauss.

Magnetometers

The variations of H, Z, and D at a given location can be measured using a magnetometer. Much of the environment in built-up areas is magnetically "noisy" as a result of human activities. Professional observatories in quiet locations, however, can very sensitively measure the fine-scale fluctuations of the local magnetic field. Results are usually displayed on a *magnetogram*, a plot of field strength or direction versus time. The magnetogram trace is indicative of the nature of any disturbance.

In its simplest form, the magnetometer need only be a suspended, free-swinging magnet which will respond to the changing local magnetic field by re-orienting itself. If these swings can be measured and recorded, the observer has a means of detecting fluctuations in the value of D, the angular deviation of the field. Such a device is the "jam-jar" magnetometer devised by Ron Livesey of the BAA Aurora Section in the early 1980s.

The jam-jar magnetometer comprises a reasonably powerful suspended bar magnet, protected from drafts by being hung on its thread inside a clear jar: despite the device's popular name, most users have found screw-top instant coffee containers best. The thread from which the magnet is suspended passes through a small hole in the center of the container's lid. A small strip of mirror is attached to the magnet, and onto this is shone a reasonably narrow (crudely "collimated") beam of light from a flashlight or other convenient source from a distance of, preferably, a few meters. The position of the reflected spot of light from the mirror on the opposite side of the room, garage, or

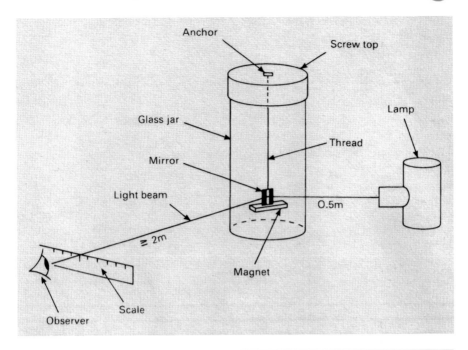

Figure 3.7. The simple, but effective 'jam-jar' magnetometer allows fluctuations in the horizontal component of Earth's magnetic field to be measured and affords the possibility of obtaining early warning of auroral activity. The free-swinging bar magnet, protected from draughts by being enclosed in a clear container, aligns itself to the magnetic field, Fluctuations can be sensitively detected by regularly measuring the position of the reflected light beam from a mirror mounted on the magnet; some refer to this arrangement as an 'optical lever'. Such devices are operated by many amateur astronomers around the word to good effect, Drawing courtesy of the jam-jar magnetometer's designer, Ron Livesey.

other convenient location for the magnetometer, is measured against a suitable scale such as a 30 cm ruler.

Readings taken at intervals will reveal the diurnal variation due to tidal effects on the ionosphere, and the occurrence of any—possibly aurora-related—magnetic activity. Operators of the jam-jar magnetometer system have picked up, clearly, crotchets, Sudden Storm Commencements, and storm/substorm activity. Early warnings of possible visible aurora have been issued to amateur observers in Scotland on the basis of magnetic field activity detected by jam-jar magnetometer operators.

It is sometimes the case that the disturbances detected using these simple but effective devices do not coincide with aurora visible at the latitude of the magnetometer operator. Apparently, magnetic disturbances at ground level can propagate to lower latitudes than the visibility of the atmospheric displays that caused them. In some respects, magnetometry of this kind offers those at lower latitudes (as in southern England, or the mid United States) their most reliable and frequent means of following the effects of solar–terrestrial interactions.

An important factor in successful operation of the simple jam-jar magnetometer is that it be located in a "quiet" environment. Spare rooms or garages are favored locations, but it seems that movement of metallic objects (garden implements in a garage, for instance) can alter the local magnetic field by a measurable extent! Likewise, the sensitivity of a well-made jam-jar magnetometer is sufficiently high that it can detect magnetic fluctuations due to passing cars or trains in the vicinity.

Once a good location has been found, the set-up should not be moved, and results thereafter will be consistent from day to day. While many operators record the deviations from jam-jar magnetometers by eye, it is possible, with a little ingenuity, to construct electronically recording versions. The best jam-jar magnetometer records may be quite readily aligned with magnetograms recorded at professional magnetic observatories using much more sophisticated equipment.

Fluctuations in the orientation of the magnet in jam-jar systems can frequently be correlated with interludes of increased activity during auroral events, as has been demonstrated on several occasions when visual and magnetometer observations have been obtained simultaneously.

More complex in its working principle and construction, but still accessible to the amateur worker in possession of electronics skills, is the *fluxgate magnetometer*, which can very sensitively detect fluctuations in the local magnetic field *strength*. Since they measure field strength rather than deviation, fluxgate systems are, apparently, better able to detect Sudden Storm Commencements than the swinging-magnet types.

Behavior of the Magnetic Field

Even under quiet solar-geomagnetic conditions, when auroral activity is confined to high latitudes, fluctuations in the magnetic field at mid-latitudes can still be detected. The Sun and Moon raise tides in the ionosphere, causing it to rise and fall slightly. Consequently, the strength of the ground-level magnetic field induced by electric currents in the ionosphere is subject to a gradual *diurnal variation*.

Increased solar activity produces a number of detectable effects. The transient increases in D-region ionization produced by solar flare-emitted X-rays, which cause Sudden Ionospheric Disturbance events (Chapter 2), are responsible for magnetic *crotchets*. During SIDs, the ionospheric conductivity rises, with consequent increases in the induced ground-level magnetic field. In common with radio SIDs, crotchets are detected only on the day-side of the Earth.

The fluctuations produced in crotchet events give a fairly regular "saw-tooth" trace on a magnetogram. Less regular are the traces produced when coronal mass ejections arrive in near-Earth space via the solar wind. Compression of the magnetosphere by the supersonic shock-wave at the CME's leading edge briefly intensifies the Earth's magnetic field, and is recorded as *Sudden Storm Commencement* (SSC). SSC events are often, but not always, followed by the development of auroral activity extending to lower latitudes.

Where SSC is followed by extensive auroral activity, the geomagnetic field is also found to be active, during the *main-phase storm*, which can last for many hours. Wild fluctuations may occur during the main-phase storm, whose effects can be particularly

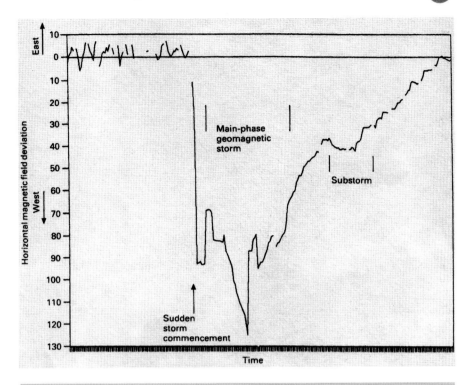

Figure 3.8. The effectiveness of the jam-jar magnetometer in recording magnetic field fluctuations is clearly shown in this magnetogram, reproduced by kind permission of the observer, Tony Michell of Beckenham, Kent, England. Mr Michell uses a personal computer to record the motions of the free-swinging magnet. This magnetogram shows the progress of a geomagnetic storm around 24–25 December 1989, when poor weather prevented visual observations from the British Isles. At left, minor fluctuations of the undisturbed field are seen. Arrival of the shock wave of a coronal mass ejection in the solar wind produced the sudden storm commencement (SSC), indicated by the arrow. The main-phase storm then set in, followed by a gradual recovery phase, on which a substorm can be seen superimposed towards the right of the magnetogram. Recovery of the magnetic field to its quiet condition took about three days from the SSC.

marked locally in the period during which the magnetic observatory is closest to the disturbed auroral oval, on Earth's night-side. Peaks of activity during which the aurora becomes coronal give rise to very sharp fluctuations in local magnetic field intensity (bringing a virtual "collapse" in severe geomagnetic storms). The main-phase storm is brought on by intense currents in the upper atmosphere, as electrons rain in from the magnetotail plasma sheet.

The *recovery phase* following the onset of a geomagnetic storm can take several days, particularly after a major CME event. Superimposed on the gradual recovery are *bays*, corresponding to substorm activity: under normal conditions, only those observatories at higher latitudes will detect the magnetic field variations due to substorms.

Amateur magnetometer operators who keep a regular eye on their equipment have the chance, then, to detect in real time the onset of disturbed geomagnetic conditions and look for possible accompanying auroral activity.

Indices of Geomagnetic Activity

Magnetometers are operated by numerous professional bodies for the purposes of monitoring the temporal fluctuations of the global magnetic field. Indices of daily activity are published by a number of these institutions, and can subsequently be correlated with auroral activity. Several different indices are published; a few of the more commonly used are listed below:

The aa Index

One useful method of measuring the degree of disturbance of the geomagnetic field is to compare activity at two stations at equivalent—antipodal—locations in either hemisphere. Such a method has been employed since 1868 by the Institut de Physique du Globe de Paris, to produce the aa index. The aa index is based on 12-hour averages of activity from antipodal stations. Other versions may adopt 3-hour averages, while the Aa index presents a 24-hour average of aa values.

The Kp Index

The University of Gottingen in Germany is the location of one of the world's earliest established magnetic observatories. Since 1932, the Institut für Geophysik at Göttingen has originated what is probably the most widely used index of geomagnetic activity, the Kp index. This index, introduced by Bartels, is based on magnetometry by 12 observatories worldwide, mostly at sub-auroral locations in the northern hemisphere. These record variations in field intensity, which are subsequently averaged for all stations over intervals of 3 hours. Kp is determined on a semi-logarithmic scale, based on the average variation in magnetic field intensity as indicated in Table 3.3.

A planetary index, Kp, of greater than 5 implies storm conditions, and can often be correlated with visible displays at lower latitudes. Summaries of the daily Kp indices for the previous month are issued monthly.

Table 3.3. The Kp index

Variation (nT = gammas)	0	5	10	20	40	70	120	200	330	>500
Kp index	0	1	2	3	4	5	6	7	8	9
3-hour ap equivalent	0	3	7	15	27	48	80	140	240	400

The Ap Index

An alternative index of geomagnetic activity, again originated from Göttingen, is the Ap index. This is similar to the aa index, in taking account of readings taken at antipodal stations, but averaged over 24 hour intervals. Unlike the roughly logarithmic Kp index, Ap is a *linear* measure of magnetometer deviation. The two indices can, however, be broadly related by the 3-hourly equivalent ap index (Table 3.3).

Radio Aurorae

Disturbances of the ionosphere (Chapter 2) during auroral conditions are something of a double-edged sword for radio communication—some forms may be enhanced, others disrupted. Amateur radio operators—many of whom are amateur astronomers, and vice versa—can sometimes obtain early warning of potential visual events under these conditions. Many local astronomy clubs maintain good links with their local radio "hams," who can on occasion give the alert to ongoing auroral conditions, which may set in during the afternoon, offering the chance of seeing a display in the following evening.

Sudden Ionospheric Disturbances resulting from solar flares are detrimental to high frequency (HF) radio communication. These events lead to fadeouts as a result of increased D-region ionization and absorption of radio waves, which occurs over the whole day-side of the Earth, irrespective of geomagnetic latitude. At high latitudes, the arrival of accelerated protons following a solar flare can give rise to polar cap absorption events, during which HF radio communication is disrupted. Enhanced ionization at the altitudes of the higher ionospheric E and F layers can also be disruptive, again as a result of increased absorption.

Enhanced ionization under auroral conditions can also be used for the *reflection* of VHF radio waves. Reflection perpendicular to auroral arc structures comes straight back to the transmitter as *backscatter*. Professional ionospheric research often involves the use of powerful radar signals, which may be detected as backscatter from auroral structures using amateur radio equipment.

Anomalous reception of signals from radio *beacons* can also indicate the onset of auroral conditions. Typically, such beacons have a range of about 300 km in all directions under normal conditions, but reflection of beacon signals from an auroral arc can alter the range in certain directions. For example, the Black Isle Beacon "went auroral" (that is, became receivable!) for radio operators in central Scotland, for whom it is normally too distant for reception, during the March 1989 storm.

Rather than aiming to detect backscatter from research transmitters or beacons, amateur radio operators using the universal 144.3 and 433 MHz calling channels usually prefer to "work" the aurora in order to obtain longer-than-normal distance (DX) contacts with other operators. Basically, the aurora can be used as a radio-reflective surface from which VHF signals can be bounced, to permit "bi-static" backscatter contacts. Visual and radio events need not overlap; for instance, radio amateurs enjoyed excellent auroral contact conditions in February 1984, at a time when low-latitude visual events were scarce.

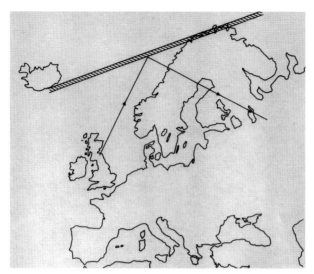

Figure 3.9. Use of an auroral arc to scatter radio signals over greater than normal distances. In this instance, an arc—aligned to the geomagnetic field lines, and lying some way north of the transmitter—is used to reflect radio signals into eastern Europe from Britain.

From a given location, there is a unique, theoretical limit—the "boundary fence"—within which bi-static auroral contacts are possible. This typically defines an oval region extending some 2000 km east–west, and 1000 km in the north–south direction. The best reflection is from discrete auroral arcs aligned to lines of geomagnetic latitude. Operators in the UK, for instance, use discrete arcs lying north of Shetland for the reflection of signals eastwards into Scandinavia, or to eastern European countries such as Hungary. The unusual situation of auroral activity *south* of the UK allowed British radio amateurs to make contacts with operators in Italy during the Great Aurora of March 1989.

Movement of, and turbulence within, the reflective auroral surface leads to interference and signal distortion. Voice and Morse transmissions become "raspy" (and, often, difficult to decipher) as a result of Doppler motions, and interference as successive radio waves overlap.

Enhanced signal reflection during auroral conditions is primarily from the clouds of ionization in the aurora itself, rather than from enhancements of the ionosphere.

Both time of day and geomagnetic activity influence the efficiency of auroral radio contacts. Under quiet geomagnetic conditions (Kp less than 5), the peak period for establishing long-distance contacts from a mid-latitude station is between 17–19 hours local time, with a secondary peak around 21 hours. Around local magnetic midnight, when the station lies closest to the Harang discontinuity, auroral radio communication conditions are very poor.

Communications can occur only if the auroral oval lies within a certain range of the transmitting and receiving stations. Such conditions are met around the times of the observed peaks. Onset of increased geomagnetic activity and expansion of the oval will alter the peak times for bi-static contacts, which are then made earlier and later: typically, under conditions of Kp greater than 5, peak times for long-distance auroral contacts are 14–19 hours and 21–00 hours local time.

Seasonal effects also operate: the secondary peak is more pronounced around the equinoxes. Summer is generally a poor time for radio aurora DX contacts.

Geomagnetic storm conditions (Kp greater than 5) also enhance the secondary peak in auroral backscatter bi-static communications.

The efficiency of auroral radio contacts also varies as a function of time in the sunspot cycle. In contrast with visual events, radio aurorae show a peak after sunspot maximum, around the time when coronal holes are common. Contact during intense storms, typical of earlier in the cycle, can be very erratic due to rapid movements within the aurora, and the development of denser clouds of ionization which may absorb even VHF signals.

Radio auroral phenomena can occur in the absence of visual events, and vice-versa. Radio events seem more frequently to be recurrent, and are thus somewhat easier to forecast.

Internet Resources

The Internet-connected modern amateur astronomer can, of course, eschew the "DIY approach" and simply tap in to data available in the public domain. For the would-be aurora observer, there are a number of extremely useful information sources, from which real-time information on prevailing solar and geomagnetic activity can be obtained.

One of the most frequently accessed sites is http://www.spaceweather.com Here can be found reports on the current state of solar activity, including sunspots and coronal holes, along with forecasts for the likelihood of aurorae at high- or mid-latitudes, Images and reports of recent auroral displays are frequently posted here, and there is a useful archive of observational material and news stretching back to 2001. For those who are interested, spaceweather.com also covers other transient phenomena such as meteor showers or close-approaching comets and asteroids This is certainly an ideal site for the newcomer.

Somewhat more advanced in its content, and extremely valuable to more experienced hands, the NOAA Space Environment Center at Boulder, Colorado, presents a wealth of technical information in its website at http://www.sec.noaa.gov. Aurora enthusiasts will find the "Today's Space Weather" and "Space Weather Now" pages particularly useful. Through the former, NOAA issues a broad-brush forecast for likely geomagnetic activity in the next 24–72 hours. Rolling graphic displays of spacecraft data provide information on recent activity including Kp values and magnetic flux conditions in the orbital environment. GOES (Geostationary Operational Environmental Satellite) data record the occurrence of solar X-ray flares, and particle counts: the onset of solar radiation storms, as fast protons ejected during solar flares arrive in near-Earth space, can also on occasion be dramatically illustrated.

The most up-to-date information is found on the "Space Weather Now" pages, including current measurements of solar wind velocity and pressure, and the intensity and orientation of the Interplanetary Magnetic Field as found by the ACE (Advanced Composition Explorer) spacecraft at the L1 position "upwind" of Earth, When the data show a strong southerly IMF and high solar wind speed the chances are that good low-latitude auroral activity is imminent or in progress, and it is well worth checking the sky. (Doing so on a largely cloudy night, when I'd otherwise not probably have bothered, led me to catch a good display of aurora through chinks in the overcast on 11–12 April 2001, for example).

The NOAA pages also offer an X-ray view of the Sun showing the positions and intensities of active regions. More detailed imagery of solar events can be found at the SOHO websites (http://sohowww.nascom.nasa.gov/ and http://sohowww.estec.esa .nl/). These include time-lapse movies from the spacecraft's C2 and C3 coronagraphs, in which approaching—potentially aurora-causing—coronal mass ejections may be evident. Hundreds of small comets, disintegrating in the inner solar atmosphere, have also been discovered in these images, the majority of them by amateur astronomers who specialize in downloading and scanning the movies.

Some of the archived coronagraph movies are well worth viewing—as an example, it is fascinating to watch the succession of CMEs around the time of the 2003 October "Hallowe'en Storms," and the near-blinding of the coronagraph by huge numbers of protons during the most active parts of the attendant intense solar radiation storm.

Both the NOAA site and Spaceweather.com will often refer to spotgroups as active regions (for example, AR 9393), rather than AAs. There is a subtle distinction, in that NOAA's numbering system for active regions includes features such as chromospheric plages, which may not, at the time, be involved with sunspots.

NOAA began numbering active regions from 5 January 1972. The regions are numbered in order of appearance, or of rotation on to the visible solar disk. Some active regions may survive an entire rotation—or, rarely, several rotations—around the Sun's averted hemisphere, and on return, a new AR designation is given—so, on its return one rotation on from its original appearance, AR 9393 was re-designated as AR 9433. In June 2001, AR 10,000 was reached and—since computers find it easier to work with four, rather than five digits—the numbering scheme reverted to 0001, and so forth. In the early 2000s, designations such as AR 972 are tacitly understood to really mean AR 10972.

In the UK, the University of Lancaster operates the Aurorawatch site (http://www. dcs.lancs.ac.uk/iono/aurorawatch/), which offers early warnings of ongoing geomagnetic activity. Initially (in 2000) run from the University of York, the pages provide real-time data—H values—from a fluxgate magnetometer at Lancaster. This system is part of SAMNET, the UK Sub-Auroral Magnetometer Network, which has a number of stations in the British Isles and overseas.

CHAPTER FOUR

Observing the Aurora

Auroral Activity at High Latitudes

Aurorae are, in the popular imagination, almost synonymous with the polar regions. Countless books have appeared over the years covering the exploration of the Arctic and Antarctic and, subsequently, on their wildlife. The aurora receives frequent mention and illustration in such texts. Clear descriptions of auroral substorm activity can be found, for example, in the writings of the late-nineteenth century Norwegian explorer Fridtjof Nansen:

> Later in the evening Hansen came down to give notice of what really was a remarkable appearance of aurora borealis. The deck was brightly illuminated by it, and reflections of its light played all over the ice. The whole sky was ablaze with it, but it was brightest in the south; high up in that direction glowed waving masses of fire. Later still Hansen came again to say that now it was quite extraordinary. No words can depict the glory that met our eyes. The glowing fire-masses had divided into glistening, many coloured bands, which were writhing and twisting across the sky both in the south and north. The rays sparkled with the purest, most crystalline rainbow colours, chiefly violet-red or carmine and the clearest green. Most frequently the rays of the arch were red at the ends, and changed higher up into sparkling green, which quite at the top turned darker, and went over into blue or violet before disappearing in the blue of the sky; or the rays in one and the same arch might change from clear red to clear green, coming and going as if driven by a storm. It was an endless phantasmagoria of sparkling colour, surpassing anything that one can dream. Sometimes the spectacle reached such a climax that one's breath was taken away; one felt that now something extraordinary must happen - at the very least the sky must fall. *Farthest North* (1897).

Notably, when at very high latitudes in the Arctic, Nansen saw auroral activity predominantly to the south (equatorwards) under quiet conditions, and a migration northwards during the active phase of the westward traveling surge (Chapter 2).

The Auroral Zones

Scientific measurements of the location of auroral features suggested, by the late 1950s, that activity at any given instant occurs principally within oval regions disposed around the geomagnetic poles in either hemisphere. The existence of these auroral ovals was confirmed by observations from satellites in high-altitude polar orbits, notably Dynamics Explorer-1. Under quiet conditions, the auroral ovals form rings of 4000–5000 km diameter around the geomagnetic poles, fixed in space above the rotating Earth.

Each oval is displaced somewhat, extending farther toward the equator at the magnetic midnight point (in a straight line from the Sun through the pole) on Earth's night side. The *auroral zones* where aurorae are most frequently seen are those regions where earth's daily rotation will carry observers close to the equatorwards boundary of the auroral oval. The northern hemisphere auroral zone, about 10 degrees wide in latitude, and centered around 65°N latitude, is a ring stretching from the North Cape of Norway, to Iceland, northern Canada, Alaska, and on to northern Russia.

Under disturbed conditions, when the Interplanetary Magnetic Field in the solar wind turns southwards, the magnetospheric circuit channelled through the auroral ovals increases in power, leading to expansion away from the poles, and a broadening of the ring of auroral activity particularly on the night-side.

The northern and southern auroral ovals are essentially mirror-images of one another, as was confirmed in 1967 by simultaneous photography from NASA aircraft flying at conjugate points (locations of equivalent geomagnetic latitude and longitude) in

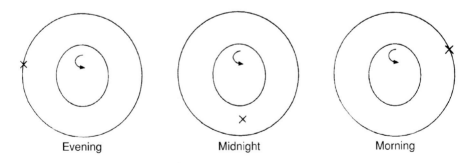

| Evening | Midnight | Morning |

Figure 4.1. An observer at a given location ('X' in the diagram) is carried closest to the position of the quiet-condition auroral oval by the Earth's rotation around magnetic midnight, being less favourably placed in early evening and early morning. The auroral ovals can be regarded as more or less fixed in space above the rotating Earth: the maximum zones of auroral frequency are those which are carried nightly under the normal, undisturbed maximum equatorwards extent of the ovals.

Table 4.1. Some Geographical Locations and Their Geomagnetic Latitudes

Location	Geographical Latitude (°N)	Geomagnetic Latitude (°N)
Los Angeles	34	45
New York	41	52
Montreal	46	56
Anchorage	61	61
London	51	54
Edinburgh	56	59
Reykjavik	64	69
Copenhagen	55	55
Paris	49	51
Madrid	40	44
Rome	42	40
Moscow	55	53

either hemisphere. Activity in the northern oval is duplicated in the southern—though satellite observations in 2004 suggest that the "mirroring" is not quite perfect!

Location, Location, Location

An observer's chances of witnessing aurorae on a regular, frequent basis depend very much on location. Obviously, those resident in the auroral zone will have the greatest chance of witnessing displays whenever skies are clear. Equatorwards from the auroral zone, clearly those at higher latitudes are likelier to benefit from any slight expansion of the auroral oval than those in mid- or low latitudes. Geomagnetic (rather than geographical) location is the key. Thanks to the offset of the north geomagnetic pole, observers in North America are currently favored over their European counterparts at equivalent geographical latitudes in terms of auroral occurrence: despite being well south of London, New York enjoys a similar average annual frequency of displays.

One of the world's most productive aurora observers (having logged more than 2000 nights of activity up to the summer of 2006!) is Jay Brausch, of Glen Ullin in North Dakota. Glen Ullin has a geographical latitude of 46°48′, but a geomagnetic latitude of 56°. Brausch frequently makes reports of aurorae at times when they are invisible to northwest European observers thanks to the summer twilight. Weather conditions are also apparently more favorable in North Dakota. In recognition of his work, Jay Brausch was awarded the Merlin Medal of the British Astronomical Association in 1993.

Table 4.1 lists the geomagnetic latitudes for some principal cities.

Polar Cusp Aurorae

While most passes around the bow-shock and on into interplanetary space without any magnetospheric interaction, a certain, very important, proportion of solar wind plasma can enter near-Earth space via the high-latitude polar cusp regions on the dayside of the auroral oval. Electrons trickling into the high atmosphere at these locations give rise to pallid, weak auroral displays, which may cover the entire sky with a diffuse background glow. Polar cusp aurorae are seen only at high latitudes during midwinter, when the noon sky is still sufficiently dark to render auroral activity visible. (By local midnight, Earth's rotation has carried the high latitude observer away from under the polar cusp.) One such location is Adventdalen (78° 17′N, 15° 55′E) on Spitsbergen Island in the Barents Sea, the site of a professional auroral observatory, the Nordlysstasjonen operated by the University of Tromso. At Adventdalen, a 6-week observing season, centered on the winter solstice, favors the recording of polar cusp aurora.

Dayside aurorae seen under undisturbed conditions in the midwinter noon at high latitudes follow a regular pattern. For a period of one to four hours before local noon, diffuse patchy green aurora, with emission predominantly at 557.7 nm, is seen, resulting from excitation of atomic oxygen by incoming electrons with energies below 500 eV arriving via the polar cleft. For the two-hour period around noon, diffuse weak (often sub-visual) red OI emissions at 630.0 and 636.4 nm, produced by low-energy (10–50 eV) electrons entering from the solar wind directly via the polar cusp, are seen. Cleft electrons then give rise to discrete green (557.7 nm emission) arcs for up to four hours following noon.

Polar Cap Absorptions (PCA)

Violent solar events can accelerate protons to energies of between 1 and 100 MeV. High-energy protons may arrive in near-Earth space via the solar wind soon (sometimes as little as four hours) after the observed flare, and are guided by the magnetic field into the polar cap regions within the auroral ovals. Here, ionization of atmospheric particles by energetic protons gives rise to polar cap absorption (PCA) events, which can last for several days. Increased electron density in the D-region of the ionosphere at 80 km altitude results in absorption of high-frequency radio waves within the auroral zones at such times.

Associated with PCA events are diffuse, weak (usually sub-visual) *polar glow aurorae*, which may fill most or all of the polar cap region. These events are most readily studied using photometry or spectroscopy. When sufficiently bright to be noticeable to the naked eye, polar glow aurora appears pink or red as a result of enhanced nitrogen and oxygen emissions at 630.0 and 636.4 nm. N_2^+ 391.4 nm emission also shows enhancement. These emissions occur at altitudes of 50–70 km (much lower than most auroral features), equivalent to the expected penetration of protons in the 5–30 MeV range.

Enhanced N_2^+ 391.4 nm emission occurs as soon as accelerated protons begin to arrive, reaching a peak some 24 hours later—coincident with the arrival of the main CME shock wave produced in the solar wind, which brings the onset of magnetic SSC about 24 hours later.

Photographic studies, aided by narrow-passband interference filters, of polar glow aurora have been carried out from Spitsbergen. The days around the major auroral storm of 8–9 February 1986 (Chapter 5), for example, were accompanied by visible polar cap aurora, possibly showing striation; such displays are more usually reported as homogeneous.

Theta Aurora

One of the surprising discoveries made by the Dynamics Explorer-1 satellite was the occurrence of "bridges" of weak auroral activity connecting across the inside of the auroral oval, and aligned to the noon–midnight line (Figure 4.2). These transpolar arcs, commonly described as "theta aurora," from the resemblance of the auroral oval to the Greek letter when they are present, appear at times when the Interplanetary Magnetic Field turns northwards, magnetospheric activity in general declines, and the auroral oval fades. Theta aurora events are accompanied by a pronounced weakening of the auroral electrojets (Chapter 2).

Under low-activity conditions, an observer at a high-latitude site, within the auroral oval, is carried under the theta aurora twice daily—around noon and midnight—by Earth's rotation. The north–south alignment of transpolar arc features contrasts with the normal east–west alignment of substorm or storm arcs in the auroral oval population. Studies of the evening sector theta aurora have been made from Spitsbergen.

Theta aurora is produced by low-energy (0.6–1.0 keV) electrons from the magnetotail. These electrons do not undergo acceleration in near-Earth space, and are therefore less penetrative than those giving rise to classical polar aurora; the base altitude of theta aurora is high, typically around 150 km.

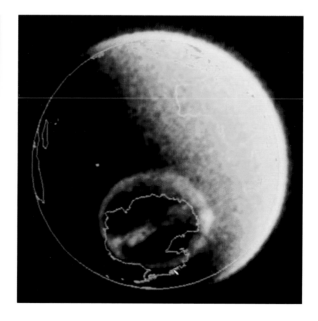

Figure 4.2. A Dynamics Explorer 1 view of the southern auroral oval from high above Antarctica, showing the oval in 'theta' configuration. Theta aurora comprises a bridge of relatively weak activity running along the noon-midnight meridian within the oval, and occurs at times of low geomagnetic activity. Image courtesy of University of Iowa.

Visually, theta aurora appears faint and relatively quiescent, comprising long, thin, more or less static rays. Ray tops may reach into the sunlit upper atmosphere, where blue–violet N_2^+ emissions at 391.4 nm and 427.8 nm are seen. Predominantly, however, theta aurora emission comes from the auroral green OI 557.7 nm line. These discrete arc structures produced in quiet times contrast with the diffuse polar glow aurora present within the polar cap at times of high geomagnetic activity.

Substorm Events

Polar cusp, PCA, and theta aurora events occur inside the auroral oval, polewards of the auroral zone. In order to witness these, the observer is required to travel to the hostile Arctic (or Antarctic) environment—so these are forms that will not, normally, be accessible to the amateur aurora-watcher. Experienced amateurs have, on occasion, been invited to carry out studies at professional observatories inside the Arctic Circle: Dr. Alastair Simmons, a long-standing contributor to the BAA's work, has spent several winter seasons at Adventdalen on Spitsbergen observing polar cusp aurorae, for example.

Rather more accessible to the average amateur are the substorm events that occur on the night-side of the auroral oval. These can be observed from locations in the auroral zone such as Tromso in Norway, or Anchorage, Alaska, which may even be set up to welcome "auroral tourists." The aurorae sometimes witnessed by passengers on transatlantic flights also usually belong to the substorm/oval night-side population.

Displays are most frequently observed from locations in the auroral zones between about 17 hours and 02 hours local time: this is the period during which Earth's rotation carries the high-latitude observer closest to the auroral oval's maximum equatorwards extent. Also, the most intense auroral activity occurs in the pre-midnight sector of the oval—the evening sector is dominated by greenish arcs, bands or rayed draperies, while the forms seen in the morning sector tend to be reddish and more fragmentary and diffuse.

Figure 4.3.
Curved, multiple band structure during a higher-latitude substorm auroral display. Image: NASA.

The pattern seen at a high-latitude site on a given night will depend on a number of factors. These include the precise time at which the disturbance leading to the substorm reaches its peak, and the overall level of geomagnetic activity. One-off chartered flights to see the aurora are equally at the mercy of these circumstances, and while good activity is usually quite likely, it is by no means guaranteed at a given time on any particular night.

Commonly, auroral activity at high latitudes will begin in the evening as quiet *arcs*, often multiple, and usually homogeneous, i.e., smooth, and lacking internal structure. The onset of increased activity may be heralded by brightening of the arcs, and the appearance of vertical *ray* structures. During the *expansive phase*, rayed arcs and folded *band* forms may rapidly extend across the whole sky, showing strong color (green rayed structures with a red lower border resulting from N_2 emission, for example) and vigorous movement. This increase in north–south extent is a consequence of overall broadening of the night-side auroral oval. Loop or spiral structures may appear as the auroral oval is contorted by the westward-traveling surge. Eastwards counter-streaming may also be seen in the rays. The narrowness of individual bands or arcs becomes apparent as these pass overhead; while rayed structures may have a vertical extent of several hundreds of kilometers, and an east–west extent of thousands of kilometers, each may be only 1 kilometer or so in width. Folded bands, as they pass overhead, will show coronal structure as a result of perspective. This activity, lasting for perhaps 30 minutes, is described as "auroral breakup."

Auroral breakup is particularly spectacular if it occurs around local midnight: the equatorwards edge of the auroral oval at its midnight point undergoes the most marked brightening. When breakup occurs in the early evening, the observer may find much of the activity in the eastern part of the sky.

As activity subsides, the auroral forms become more diffuse and fade. Such aurora is common in the morning sector of the oval. Pulsating forms mark the recovery phase of the substorm. Even at times of low activity, following substorms, the sky may be suffused by a faint background of auroral light.

Following breakup, and the recovery phase, quiescent arcs may re-form, prior to the onset of a further round of activity. Under intensely disturbed geomagnetic conditions resulting from violent sunspot activity, substorms may follow closely one upon another, leading to hugely impressive aurorae maintained through much of the night. On the other hand, when geomagnetic activity is low, substorms may comprise little more than a brightening of the arc structures in the evening sector, with no ensuing major breakup.

Those traveling to higher latitudes to observe substorm aurora will usually wish to obtain a permanent record—either by digital camera or on photographic emulsion. Among the latter, Kodak Elite (Ektachrome) and Fujichrome slide films are long-established favorites with aurora photographers. The basic guidelines covered later with respect to imaging lower-latitude aurorae apply—use a tripod to steadily mount the camera, and a locking cable release to give shake-free shutter operation. Sadly, attempts at imaging from aircraft are often prone to unavoidable camera shake—planes can be quite unstable observing platforms, and the cleanliness or otherwise of the available window may also be an issue.

Fast films (ISO 400) or equivalent sensitivity setting for a digital SLR, are preferable. If observing in very cold conditions, bear in mind that films can become brittle and may snap inside the camera if wound on with excessive vigor. Batteries—if anything, even more of a consideration for digital cameras—run down quickly in the cold: a warm pocketful of spares is always worth carrying!

A wide-angle lens (28 mm or similar, for 35 mm format cameras) at $f/2.8$ or faster is ideal for catching large areas of aurora-filled sky. Exposure times during the intense activity of auroral breakup will be a good deal shorter than one might use to record displays at lower latitudes. Bright, rapidly moving substorm aurora can be adequately captured in exposures of as little as 1–2 seconds at ISO 400: longer exposures run the risk of being washed out, and the detail blurred by movement of the auroral forms.

Geomagnetic storms

Although the initial stages may show an activity pattern similar to the onset of substorms—including brightening of forms, and development of a westwards-traveling surge—major geomagnetic storms can also carry the aurora equatorwards *away* from high-latitude skies. It is documented, for example, that during the great mid-latitude storm of 25 January 1938 when observers in southern Europe were able to see striking red aurora across much of the sky, those in Tromso had no aurora whatsoever!

Auroral Activity at Lower Latitudes

While displays can be an almost nightly occurrence for those living in the higher-latitude auroral zones, the aurora is a rather less frequent visitor to the skies of the majority of us who live in middle latitudes. These might reasonably be defined as the latitudes of the central United States, northern Europe (including the British Isles) south of Scandinavia, or Australasia. Under normal conditions, when the solar wind is steady and no major geomagnetic disturbance is in progress, the auroral oval lies well polewards from observers in these latitudes, and even vigorous substorm activity is out of range. As described in Chapter 2, however, the arrival of pockets of intensified magnetic field and increased electron density associated with coronal mass ejections can lead to brightening and broadening of the auroral ovals on the night-side of the Earth: under such conditions, the aurora may be seen at lower latitudes.

The frequency with which aurorae can be seen is, as already described, strongly dependent on geomagnetic latitude. As a rough guide, a study of more than 30 years' auroral reports from UK-based observers, conducted by Ron Livesey (until 2005, Director of the BAA Aurora Section) suggests an average frequency of displays of 5 per annum in the south of England (geomagnetic latitude 53°N), reaching 60–70 per annum in the Moray Firth in northeast Scotland (geomagnetic latitude 59°N). The Moray Firth is notable as one of the best locations from which to see aurora in the British Isles, being blessed with dark skies and a good proportion of clear nights.

Factors such as weather and summer twilight have to be taken into consideration, of course. Perhaps problematical to most observers is the spread of light pollution: of those five displays a year potentially visible from London, the majority will probably be relatively minor events, lost in the foreground glare. From my own regular observing location in Sussex in the south of England, I am certainly wary of any glow amidst the other light pervading the northern sky on some nights: unless it's obviously structured, colored or shows characteristic auroral motion and/or changes in brightness (described below), I would hesitate to record light in my northern sky as aurora! Very major displays associated with geomagnetic storms are, however, unmistakable—there are reports of such events as the April 2000 storm being visible even from town center locations.

While no two displays are exactly alike, aurorae at mid-latitudes tend to follow a fairly set activity pattern, often building to a climax then falling back to the polewards horizon. The most major storms may show multiple cycles of build and decline. Models suggest that penetration of solar wind plasma into, and subsequent stress on, the magnetosphere may be either continuous or spasmodic. If the first holds, then moment-to-moment changes in auroral activity could be taken as a direct reflection of the continually changing Interplanetary Magnetic Field. Alternatively, spasmodically entering solar wind plasma may accumulate in the magnetotail until sufficient stress is generated, then released in a huge earthwards surge of accelerated particles, giving rise to a short interval of enhanced auroral activity. This latter mechanism, working on a 40–60 minute cycle, could account for the semiregular outbursts seen during major storms.

Geomagnetic Storms

The most spectacular aurorae at mid-latitudes are certainly those which follow the arrival, in near-Earth space, of energetic plasma and strong magnetic fields associated with coronal mass ejections. CMEs are usually connected with large, actively changing sunspot groups, which are often commonest in the early parts of the cycle.

A visual auroral display resulting from a solar flare event will often begin in early evening as a quiet *glow*, low over the polewards horizon (Figure 4.4). Clouds can often appear in silhouette against the glow, while stars can be seen quite clearly through it. When the aurora has this appearance, it is easy to understand how it came to be called the "northern dawn."

If the geomagnetic disturbance is relatively small, producing only a slight expansion of the auroral ovals, such a glow might represent the maximum extent of the auroral display: obviously, observers at higher geomagnetic latitudes will enjoy a more extensive display under such circumstances, while the lower latitude observer merely witnesses the uppermost parts. Glows are often faint, sometimes little more obvious than the milky way, and will often be missed, especially in areas badly afflicted by artificial light pollution, or in conditions of bright moonlight.

Observers must be cautious to avoid misidentifying glows produced by artificial lights in towns or villages along the line of sight to the pole. Especially on slightly hazy nights, spurious auroral reports due to light pollution are not uncommon.

Figure 4.4. Auroral glow (G). From lower latitudes, particularly, this may be all that is visible in many displays, representing the uppermost parts of activity which may be more structured and impressive at higher latitudes, The green oxygen emission seen here often appears more bluish to the eye; films usually enhance the colour of auroral emissions. Occasionally, a glow of this type may be the precursor to increased activity later in the night. Image: Dr Dave Gavine.

Auroral glows may fade away after a time, without anything further being seen. On occasions when the aurora is likely to become more active, however, the glow may brighten and rise higher into the sky. From this stage, a more definite structure may develop in the aurora. The glow often resolves itself into an *arc* structure spanning the poleward sky (Figure 4.5). The arc, which has been likened by some observers to a white or pale green rainbow, will usually show no internal structure in the early part of the display and in such a condition is described as *homogeneous*.

The highest point on the base of an arc lies roughly in the direction of the observer's magnetic pole. Thus, for observers in the British Isles, for example, the highest point on auroral arcs is usually a little to the west of north. Reflecting the sharp lower boundary for auroral emissions in the high atmosphere, the base of the arc is much more clearly defined than the often rather diffuse upper extremities.

After a time, a homogeneous arc will often develop bright areas along its length, from which vertical *ray* structures appear (Figure 4.6). *Fading* of the arc may also be the prelude to ray formation. The rays can be fairly static, though in an active display some degree of movement is more normal. On occasion, the observer may see only isolated bundles of rays, whose appearance can be likened to searchlight beams. In a brightening display, color may become more evident, and the differing red and green oxygen emissions along the vertical extent of the rays can sometimes be very striking:

Figure 4.5. Homogeneous arc (HA) spanning the northern sky from Ayrshire on 22–23 February 1988. Glows like that shown in Figure 4.4 may sometimes resolve into such structure, which again may be a precursor to more extensive activity as the night progresses and storm activity develops. Arcs like this are also common during intervals when geomagnetic activity is enhanced during immersion in a coronal hole stream. Image: Tom McEwan.

green emissions predominate low down, red higher up (Chapter 2). By this stage, the aurora may be quite obvious, even from light-polluted locations.

Folding of the arc upon itself produces a ribbon-like *band* (Figure 4.7), which again may be homogeneous or rayed. Rapid movement of rays along the length of a band produces the "curtain" effect, commonly illustrated in popular astronomy books. This effect is not an illusion, but is indicative of the fact that the long auroral rays (perhaps stretching upwards for hundreds of kilometers) rise from a narrow band whose latitudinal extent may only be a kilometer or so.

Rarely, at the very climax of an extremely violent auroral display, activity may pass overhead and into the equatorwards half of the observer's sky. At this stage, the rays and other features will appear to converge on a single area of the sky as a result of perspective, producing the form of a *corona* (Figure 4.10). The corona is normally centered, at mid-latitudes, some degrees equatorwards of the true zenith in the observer's sky, near the *magnetic zenith.*

In a typical storm, the corona—if it forms at all—is usually short-lived, particularly at lower latitudes. Auroral activity will normally fall back toward the polewards horizon within a few minutes. There, the aurora may remain as a glow for a time before fading away.

Figure 4.6. As activity increases, a homogeneous arc may develop vertical ray structure along its length, producing a Rayed Arc (RA) In this instance, photographed from Edinburgh, the rays are fairly short (R_1A). Image: Dr Dave Gavine.

Figure 4.7. In more active displays, the arc will rise higher into the sky and take on a more ribbon-like appearance, as a band. This rayed band (R_2B) imaged by the Author also shows the variation in predominant oxygen emission with altitude in the display: green 557.7 nm emission predominates in the lower parts of the aurora, whilst the red (630.0 nm) emission is found at higher altitudes, at the ray-tops.

Figure 4.8. Between late spring and early autumn from higher latitudes, long auroral rays may extend into sunlight, resulting in the purple resonance emission of molecular nitrogen, as seen here in this display from Edinburgh on 25–26 April 1990. Image: Dr Dave Gavine.

Figure 4.9. Blue-purple nitrogen emissions predominate in this strong auroral display photographed from Edinburgh on 28–29 July 1990. Multiple rayed bands are present, covering much of the northern sky. As for Figure 4.8, a semi-fisheye lens was used to record the image, resulting in the apparent distortion of the horizon. Image: Dr Dave Gavine.

Figure 4.10. Undoubtedly the most impressive form the aurora can take during a major geomagnetic storm is that of the Corona (in this instance R_3C). Rays and other features appear, as a result of perspective, to converge on the observer's magnetic zenith. This striking example, framed against the stars of Cygnus and Lyra, was photographed from Elgin in northern Scotland on 21–22 October 1989. Image: Richard Pearce.

Sometimes, especially during a particularly violent storm, the aurora may continue to show rayed activity, rising to coronal peaks several times in the course of the night. The intense aurora of 13–14 March 1989 was remarkable in showing coronal forms more or less throughout the night at higher temperate latitudes. It is also possible for an aurora, which has fallen back to a glow, to go through the entire sequence from glow through rayed to coronal forms several times on an active night.

Other forms of aurora can accompany the arcs, bands, and rays. The sky may be suffused by a weak background *veil* during some displays, and *patches* (or surfaces) of homogeneous auroral light outlying the more structured parts of the display can also be seen (Figure 4.11). Unusual aurorae comprised only of diffuse patches have been reported, an example being the display witnessed by British observers on the night of 29–30 August 1978: such events might frequently be the source of UFO reports.

Mid-latitude aurorae are subject to variations that can be of long or short timescales. Arcs, bands, and patches often exhibit slow *pulsations* over the course of several minutes. *Flaming* activity, in which waves of light sweep rapidly upwards from the horizon, locally brightening features as they pass, is much more rapid, with several waves per

Figure 4.11. On occasion, the aurora may appear as discrete patches (sometimes called surfaces), in this case HP. This patch of auroral light, brightening and fading over timescales of several minutes, appeared to be part of a incomplete arc structure, during an unusual but widely-observed display seen over Scotland on 27–28 August 1978. The photograph, showing the aurora against the stars of Bootes and Corona Borealis, was taken from Fort Augustus in the Scottish Highlands. Image: Dr Dave Gavine.

second at its most intense. Though most often seen in the declining phase of an aurora, flaming sometimes precedes corona formation. *Flashes* resembling lightning are sometimes reported, while intense aurorae may show *flickering* as light passes horizontally through features.

Movement also takes place over differing timescales. Quiescent features may shift their position in the sky only very slowly and subtly. Rays are often seen to drift along the length of arcs and bands. A particularly eerie effect—it has certainly made the hairs on the back of my neck rise when seen from remote locations in Scotland—of the aurora is the slow, silent drift of isolated ray bundles, resembling searchlight beams, above hills or other features on the polewards horizon, as viewed from a remote, dark location. Folding and rippling of a rayed band strengthens the appearance of "curtain" structure within an auroral display. Rapid movement of rays along an arc or band produces the effect of streaming. Coronal rays can show a "cartwheel" rotation around the centre from which they appear to radiate.

The vigorous aurorae at mid-latitudes, which may follow solar flare events, tend to last only for one or two nights at most. There will usually be a single night of particularly intense activity, perhaps flanked by nights of less major aurora. Unlike coronal hole aurorae, flare events tend to be seen only at a single solar rotation;

Figure 4.12. Isolated rays, or bundles of rays—outliers from a display that may be more extensive at higher latitudes—are occasionally recorded. This example shows a moderately-long single ray (R_2R) against the stars of Ursa Major. Image: Neil Bone.

Figure 4.13. Sometimes, when the aurora is visible through cloud, it is impossible to identify the structure. In such cases, it is still worth recording the presence of auroral light as N (not identifiable). This photograph captures the bright display of 11–12 April 2001, as seen by the Author from Chichester, West Sussex in southern England. Observers in France and the Netherlands had much more impressive views under clear skies.

recurrences are quite rare, since the active sunspot groups above which flares develop have, themselves, fairly short lifetimes.

Coronal Hole Aurorae

In contrast with the often violent, active events of the rising years of the sunspot cycle, those few aurorae seen from mid-latitudes late in the cycle tend to be relatively quiet. These late-cycle events are normally produced by the passage of broad, persistent coronal hole particle streams, which open up particularly a couple of years prior to sunspot minimum.

Coronal holes may persist for months on end, and an individual particle stream may sweep over the Earth several times, resulting in auroral displays which recur at 27-day intervals. Such recurrences might be more reliably forecast than the capricious flare ejections from sunspot groups. For example, a persistent coronal hole was the source of a long series of recurrent mid-latitude aurorae throughout the second half of 1985.

As with CME-associated aurorae, those events produced by coronal hole streams may often commence as a featureless glow on the poleward horizon. Activity may rise higher into the sky, assuming the form of arcs or bands, with occasional rayed outbursts. Movement and changes of brightness tend to be less marked in coronal hole aurorae, and it is not unusual for such activity to progress no further than the glow or quiet arc stage.

The expansion of the auroral ovals resulting from passage of a coronal hole stream is typically smaller than that which follows arrival of solar flare-derived particles. The relatively quiescent coronal hole aurorae are therefore usually best seen from higher latitudes; Ron Livesey rather appositely describes these as "Scottish aurorae"— coronal hole disturbances are rarely sufficiently strong to push the aurora much farther southwards than the latitude of central Scotland. Coronal hole aurorae probably figure far less frequently than flare aurorae in the historical records from lower latitudes.

"Flash Aurora"

Among reports collected by the BAA Aurora Section are a number of instances where observers record isolated, short-lived bursts of auroral activity. The duration of these can be very brief indeed—for example, one event, reported by experienced Canadian observer Todd Lohvinenko from his location in Winnipeg in 1987, lasted only five seconds, but passed through the zenith. Similar reports, albeit of less extensive activity, have been made by the vastly experienced Scottish observers Alastair Simmons and Dave Gavine, lending support to the authenticity of these occurrences. Magnetometer records may also lend support: some of these "flash aurora" events can be correlated with minor magnetic field fluctuations.

While further observations are required to absolutely verify the occurrence of these transient events, professional geophysicists have already offered explanations for them. One suggestion is that flash aurorae are correlated with high-latitude activity, and follow substorms by about four hours. Reconnection processes in the magnetotail have been suggested as the causative mechanism.

Visual Observation

The advent of spacecraft monitoring of the auroral ovals from above has undoubtedly improved the precision with which the global pattern of auroral storms can be followed. Some value is still placed on visual reports, however, and these are collected by a number of organizations, such as the Aurora Sections of the British Astronomical Association and Royal Astronomical Society of New Zealand. Modern visual observations may be compared with those in the archives, for example, allowing their correlation with data from pre-satellite times. Such work can reveal whether auroral behavior has changed significantly in historical times. Annual summaries of auroral observations are compiled and published by the BAA Aurora Section in the Association's *Journal*.

It is to the advantage of the observer that the aurora shows a fairly small range of typical forms. A standard reporting code has been in use for many years, allowing the state of auroral activity at any given moment to be accurately and concisely noted in a form useful to those wishing to derive positional and other information. The International Union of Geodesy and Geophysics' 1963 *International Auroral Atlas* provides photographic examples of the aurora's typical appearance.

Visual observers submit reports, accurate to the nearest minute or so, of the form(s) of aurora present on a given night. Times should be given in Universal Time (UT; equivalent to GMT, and used as a standard by astronomers worldwide). Dates should always be quoted in double date format—for example 13–14 April 2006 to avoid ambiguity. The extent of auroral forms present should be given in altitude and azimuth. Azimuthal extent is estimated in degrees, with 0 due north, 090 due east, and so on. The altitude in the observer's sky of the highest point on the base of a feature such as an arc or band, denoted by the symbol h, is particularly useful. Since aurorae typically have a fairly sharply defined lower boundary 100 km above the Earth's surface, h and the observer's known latitude and longitude may be combined to provide two elements of a Pythagorean triangle, and the latitudinal extent of an auroral feature can therefore be easily deduced. Combination of several such reports, obtained simultaneously, affords greater accuracy in this method of triangulation. As a rough guide, an arc or band whose base appears at a maximum elevation (h) of 30° will be overhead at a location about 175 km polewards from the observing site.

A second altitude measurement, denoted by the symbol $\nearrow$, denotes the upper edge of an auroral feature. This is sometimes difficult to determine, as the tops of rays and other features are often diffuse and difficult to discern. Figure 4.14 summarizes the positional measurements which may usefully be recorded by the visual observer.

It is usually sufficient to make note of the condition of an auroral display only as and when it changes. There is little point in making estimates of altitude and azimuth of a quiescent arc at intervals more frequent than about 15 minutes. Rapidly changing geomagnetic storm aurorae present a challenge to the observer, and will in

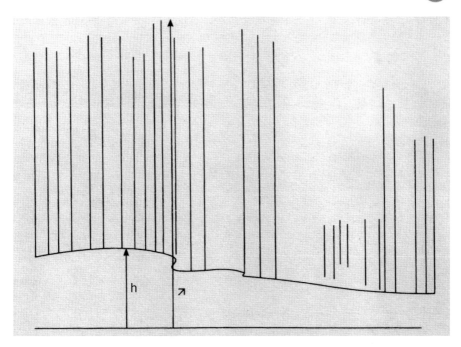

Figure 4.14. Schematic representation of the useful altitude and azimuth measurements which may be recorded by a visual observer during an auroral display (in this case for a Rayed Band).

all probability change too quickly for anything but the broad pattern of activity to be noted. Sometimes, the best way to get the details down is to make a quick, rough annotated sketch.

Changes in brightness can be recorded, and may reflect impending changes in activity. The onset of flaming, for example, may precede corona formation in a major storm, while the brightening or fading of an arc may indicate the imminence of increasing activity. A four-point scale is used to denote brightness (Table 4.2).

The accepted standard code used to concisely describe auroral forms and their state of activity is listed in Table 4.3.

Table 4.4 and 4.5 give actual examples of the code's usage from my observing log.

Table 4.2. The Auroral Brightness Scale

(i) Weak, comparable in intensity to the Milky Way.

(ii) Comparable to moonlit cirrus clouds.

(iii) Quite strong aurora, comparable in brightness to moonlit cumulus clouds.

(iv) Stronger than (iii), possibly even bright enough to cast shadows. Aurorae this bright are relatively rare at mid-latitudes, but have been reported during great storms such as those in 1989 or 2003.

Table 4.3. Standard Code for Auroral Recording

Condition   Aurorae may be either quiet (q) with no movement, or active (a).
Activity may take several forms, denoted by subscripts.

a_1 Folding of bands.

a_2 Rapid change of shape in lower structure.

a_3 Rapid horizontal movement of rays.

a_4 Fading of forms, with rapid replacement by others.

Changes of brightness are also seen, described by p followed by a subscript.

p_1 Slow pulsing.

p_2 Flaming, waves of light passing vertically through display.

p_3 Flickering; rapid, irregular variations.

p_4 Streaming; irregular horizontal variations in homogeneous forms.

Form   The aurora shows a range of discrete structures. These may be homogeneous (H), lacking internal structure, or rayed (R), showing vertical structure of varying lengths. Subscripts denote the length of rays from R_1 (short) to R_3 (long). Features may be multiple (m), fragmentary (f), or coronal (c).

G **Glow** with no other structure, often lying low above poleward horizon.

A **Arc** structure; an arch of light spanning east–west across the sky. May homogeneous (HA) or rayed (RA).

B **Band.** Folded, ribbon-like structure, often developing from an arc. May have HB or RB.

R **Rays** may sometimes be seen in isolation (eg R_1R), or in bundles (e.g., mR_2R) when no other aurora is present.

V **Veil**; a background which sometimes pervades the sky during auroral displays.

P **Patch**. Discrete area (sometimes described as a "surface") of auroral light. May appear as HP or RP.

N **Not identifiable**—may refer to auroral light seen through cloud, for example.

Colour   Characteristic colors exhibited by the aurora may be described by the use of lowercase letters at the end of feature descriptions.

a Red in upper part of aurora only.

b Red in lower border of aurora only.

c Green, white or yellow.

d Red.

e Red and green together.

f Blue or purple.

Examples: qHA1c describes a quiet (q) homogeneous arc (HA) of brightness 1, white in color; $a_3p_2mR_2B3e$ describes a bright red and green multiple rayed band, with medium-length rays showing horizontal movement and flaming activity.

Table 4.4.

Observer: N. M. Bone
Location: Campbeltown, Scotland Lat. 55°25'N Long 5°36'W
Date: 10–11 April 1982

UT	Forms	h	↗	Azimuth	Remarks
2126	qHA2c	15°	25°	270–060	Dark sky below.
	V1		30–35°		
2132	p₁HB	20°			
2135	mR₁R1-2c		20°	320–330	Short-lived.
2147	aR₁A1-3	5°		320–005	Very sharp rays.
	R₂R				To west of arc.
2155	R₂B				
2159	a₃mR₃B3e	22°	90°		Rayed forms, across entire northern sky, starting to converge at tops.
2213	aR₃B2-3	10°	60°	260–000	Very dark below.
2222	R₁R			260–355	Becoming quieter, bulk of activity in west.
2232	aR₃B3e	10°	55°	280–060	
2235	aR₂A	15°	25°	280–060	Quieter.
2243	R₂A1-2	8°	35°	280–060	Much quieter. Glare from Moon now becoming a problem.
2340	G2		20°	000	Auroral light still present. End of observations.

A minimum of equipment is required to make standard visual observations of the aurora. The naked eye suffices, and optical equipment such as binoculars will serve merely to diffuse the structure of any aurora present. The accuracy of visual altitude estimates may be improved by use of an alidade, which, in its simplest form, consists of a protractor attached to a plumb line and some sort of sighting device (Fig. 4.15). When the auroral feature to be measured is observed through the sighting device, a direct measure of altitude relative to the horizon is obtained by reading the angle of the plumb line against the protractor's scale.

Rough measurements of angular extent can be made using the outstretched hand at arm's length. For the average person, the distance between the tip of the little finger

Table 4.5.

Observer: N. M. Bone
Location: Apuldram, England Lat 50°44'N Long 0°48'W
Date: 29–30 October 2003

UT	Forms	h	↗	Azimuth	Remarks
0050	mR₂Rd	35°		320	Pinkish ray in Cygnus.
	V	25°		300–030	Pink-topped veil across northern sky.
0100	N2-3	20°		330–030	Cloud in foreground to northeast.
0103	mR₃Rc	55°		330	Long, white rays.
0108	N2-3f	25°			
0117	mR₂R2c	25°		340–000	
0118	mR₂R3d	25°		340–000	Brightened, became reddish, then disappeared!
0120	N2-3	20°			Strong at 350 azi.
0125	N2	20°		340–020	Clouding up again!

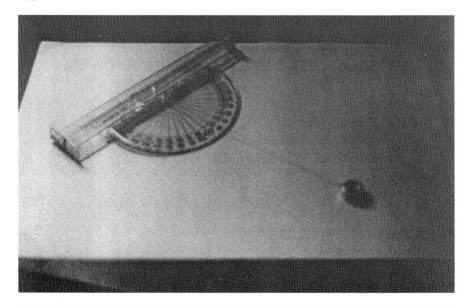

Figure 4.15. A simple alidade, basically consisting of a plumb line and protractor scale aligned with a rudimentary sighting device allows reasonably accurate determination of angular elevations of auroral features and other objects in the sky.

and tip of the thumb in the fully extended hand corresponds to an angular distance against the sky of 20 degrees.

Photography

Visual observations of the aurora can reveal patterns of activity and approximate positional information during displays. More accurate, permanent positional records can readily be obtained through photography, for which even small cameras are adequate. Only simple equipment, widely available in the amateur astronomical community, is required. The results can often be extremely aesthetically pleasing.

As in all other areas, it seems that the digital camera "revolution" has taken astronomical imaging by storm! Initially, the impression was that some areas—notably aurora and meteor photography—might remain the domain of conventional film, but advances in the sensitivity (equivalent to film "speed") in digital camera chips mean that even quite faint subjects may now be adequately recorded. Higher-end digital SLRs are certainly capable of recording aurorae. There may, however, be some issues with color rendition I have seen some very garish results from digital imaging of aurorae. These can be addressed, of course, by sensible application of "digital darkroom" software (Adobe Photoshop and Paintshop Pro are established favorites with amateur astronomers).

Film-driven 35 mm SLR cameras, until so recently in widespread use, remain eminently suitable for auroral photography, provided these have the facility for taking time exposures in excess of a couple of seconds via either a "B" or "T" setting, preferably operated by a cable release mechanism. The camera should also, ideally, be firmly mounted on a tripod.

There are few hard-and-fast rules for auroral photography. Most observers prefer to use fast color slide films, typically of ISO 400. Kodak and Fuji emulsions have been used with considerable success, and seem less prone to the garish color enhancements produced by other films. It is certainly a common characteristic of photographic emulsions that they will bring out auroral colors more marked than those visible to the eye, and greens and reds are often particularly enhanced. Most observers prefer to use the camera at full aperture, preferably $f/2.8$ or faster. Exposure time depends on the brightness of the auroral forms to be photographed.

Using ISO 400 film, weak horizon glows will be photographed with exposures of 30–60 seconds at $f/2.8$. Moderately bright auroral structures will record in 20–30 seconds at $f/2.8$, while the shutter need only be opened for 5–10 seconds at $f/2.8$ during bright, actively moving aurorae. Indeed, in the latter case, detailed structure will be lost due to auroral movement if the shutter is left open too long.

Perhaps the best advice which can be offered to those wishing to photograph, or digitally image, the aurora is to take plenty of bracketed exposures. Particularly at lower latitudes, it may be some time before a further opportunity to photograph the aurora arises, and the author would therefore suggest that observers in such locations as the south of England or the southern United States should not skimp on film or digital memory when a good display does occur!

Remember to record the time of each exposure (as for visual records, to the nearest minute in UT should suffice); digital cameras will do this automatically.

The sensitivity of photographic emulsions is often such that sub-visual aurorae can be recorded. During a quiet interlude in the November 1991 storm, for example, the author succeeded in recording diffuse red auroral emission over much of the northern sky from Sussex in the south of England using a 2-minute exposure at $f/2.8$ on ISO 400 film, although no aurora was evident to the naked eye.

CHAPTER FIVE

Historical Aurorae and More Recent Events

Aurorae in Classical Times

It is extremely unfortunate that the spread of artificial light pollution, accompanying population growth, has decreased mankind's general awareness of astronomical phenomena, including the aurora. Few city dwellers have ever seen the Milky Way, for example, and in many locations in the civilized world, only the very brightest stars and planets may be seen. In built-up areas, all but the very brightest aurorae will be swamped by the all-pervasive background sky glow. This was not always the case, however, and historical records show that the aurora was a phenomenon well-known in antiquity when skies were darker.

It seems likely that the ancient Greeks were aware of the occasional occurrence of aurorae, which may appear in Aristotle's *Meteorologica* from 340 BC as glowing clouds. Aristotle classed these together with meteors and comets as atmospheric phenomena, believing their origin to lie in friction between hot exudates from the Earth and the innermost of the heavenly spheres.

Another example of auroral activity from classical times is a frequently related tale of the Roman Emperor Tiberius, who in 37 AD dispatched a garrison to the aid of the port of Ostia, which was thought to be ablaze when a red glow was seen in the sky to the north of Rome. It appears that the soldiers spent a long night marching towards an active auroral display!

Auroral Records from the Dark and Middle Ages

Searches through historical records from the Far East (which were often kept for astrological, rather than astronomical, purposes) by a number of workers have provided clues to the past activity of several of the annual meteor showers. Accounts of nights when, for example, "more than 100 meteors flew thither in the morning" or "countless large and small meteors flew from evening till morning" provide an insight that the Perseids, currently one of the most consistent annual showers, were active as long ago as 36 AD. The evolution of some meteor showers can be traced through the broad changes in their behavior that have occurred over the centuries. The Taurids, for example, appear to have been richer than even the modern Perseids during the Middle Ages, although they are now a weak shower, producing only very low rates. This depletion over time is consistent with existing theories on the loss of material from meteor streams due to planetary perturbations and interactions with solar radiation.

The aurora, too, may be also be found reported in historical annals, and such records have been used by Eddy and others as tracers of past sunspot activity.

Typical records of auroral activity from more southerly parts of Europe in the Dark and Middle Ages refer to battles fought across the sky with fiery swords or lances, and ships sailing in the heavens. The intense crimson of oxygen emission which frequently manifests in active mid-latitude aurorae was often equated by contemporary witnesses with blood shed over the firmament. David Gavine, a student of the astronomical history of Scotland, has unearthed some remarkable examples:

93 AD: Mony birnand spieris apperit, shottand in Ye air. Ane grete part of Callandair Wode semyt birnand all nicht, and na thing appering Yerof in Ye day. Ane grete noumer of schippis wer seen in Ye air.

352 AD: In the nicht apperit mony swerdis and wappinis birnand in Ye aire. At last thai ran all togidder in ane grete bleiss, and evanyst out of sicht.

839 AD: Offt times wes sene in Ye nicht ane fyry ordinance of armit men rusching togidder with speres in Ye air, and quhen Ye tane of thame was winscust, Ye tothir sone evanist.

Each of these was chronicled by Hector Boece (pronounced "Boyce"), the first Principal of King's College, Aberdeen University. Boece (1465–1536) produced his *Scotorum Historia ab Ilius Gentis Origine* in 1526. Some of the reported incidents appear to be records of auroral activity. The time around the turn of the eleventh to twelfth centuries AD appears to have been particularly rich for aurorae in mid-latitudes. Chinese records of naked-eye sunspots appear to corroborate this, such spot-groups being typically the most productive of extensive auroral displays.

Chinese and Korean records of apparent auroral activity from this period may also be found:

1141 AD: "...at night, a red vapour appeared on the night sky, then two other strips of white vapour penetrating through the north pole and vicinity appeared also, sometimes they disappeared and then reappeared again."

Korean records often describe the aurora as a "fire-like vapour."

Further examples of auroral records may be found elsewhere in European medieval chronicles. The 12th century Anglo Saxon Chronicle contains references to aurorae, meteors, comets, and other celestial events. Among these accounts are to be found references to the sky burning, and dire forewarnings over Northumbria—fiery dragons in the air. Further English records of aurorae may also be found in Raphael Holinshed's Chronicle (published in 1577, and apparently used as a major source of information by Shakespeare in writing such historical plays as Macbeth):

1235 AD: In North England, appeared coming forth of the earth companies of armed men on horseback, with spear, shield, sword and benners displayed, in sundry forms and shapes, riding in order of battle and encountering together there. The people of the country beheld them afar off, with great wonder. (Seen for several days.)

1254 AD: Seen by the monks of St. Albans: in a clear night . . . there appeared in the element the perfect form and likeness of a mighty great ship . . . at length it seemed as the boards and joists thereof had gone in sunder, and so it vanished away.

Anyone who witnessed the awesome auroral storm covering much of the globe in March 1989 from a reasonably dark location away from artificial light pollution will have little trouble in understanding how the then unexplained phenomenon could inspire terror in the Dark Ages mind. It must have seemed, on occasion, that the World's end was imminent as the heavens blazed!

Aurorae continued to be seen through the period right up to the invention of the telescope. Further striking examples may be found of accounts in the literature well into the sixteenth century:

1529 AD: In the moneth August was seine vpon the mountaines of Striuiling afore the sone ryseng lyk fyrie candles streimes of fyre spouting furth, in the air als war sene men in harness courageouslie inuading ilk other, and sik wondiris, quhilkes with terrable feir opprest the myndes of mony.

History of Scotland (translated) Bishop John Leslie.
Even as late as the sixteenth century, the aurora still inspired terror and thoughts of war and bloodshed in its witnesses.

As has already been discussed in Chapter 2, sunspots and aurorae became rare for a time—the Maunder Minimum—before what we recognize as the "normal" activity pattern became established. The aurorae seen by Halley in 1716 and 1719 were followed, for example, by a sighting of coronal aurora from France by de Mairan in 1726. This observation was represented on what appears to be the first "gnomonic" all-sky summary chart of auroral activity. Scientific interest in the aurora was certainly great during the eighteenth century following the end of the Maunder Minimum, and several records were made of displays over northern Europe at this time.

The naturalist Gilbert White (1720–1793) was an assiduous observer of many phenomena from his location in rural Hampshire in the south of England, as published in his *Natural History of Selborne*. White made a number of auroral observations, recorded in his Journals between 1768 and 1793. At latitude 51°N, White would have witnessed only major storms. Among his records are such examples as:

25 October 1769: A vivid aurora borealis, which like a broad belt stretched across the welkin from East to West. This extraordinary phenomenon was seen the same night from Gibraltar.

18 January 1770: Vast aurora: a red fiery broad belt from E to W.

15 February 1779: A vivid aurora: a red belt from East to West.

13 October 1787: The aurora was very red and aweful.

The Nineteenth Century

Solar and auroral activity had by the beginning of the nineteenth century long settled into the patterns familiar to modern observers. Some very high sunspot maxima occurred, notably in 1836 and 1870. As proved to be the pattern in the twentieth century, the most vigorous, extensive aurorae tended to be those in the pre-sunspot-maximum phase, an example being a fine display extending to southern England in 1847.

A major event in auroral history was the observation at 11:20 am on 1 September 1859 of a solar flare by the English astronomer Richard Carrington (1826–1875) from his observatory at Redhill, Surrey. Carrington was surprised to see a brightening within a group of sunspots that he was sketching at the time. Such "white light" solar flares are extremely rare: flares are most commonly recorded using equipment that allows the Sun to be observed in the light of hydrogen-alpha, and which had yet to be developed in Carrington's day.

Carrington's flare was followed, a day or so later, by intense auroral activity over much of the Earth. The aurora was even visible in the tropics, from Honolulu, Hawaii (latitude 21°N). Activity persisted for several days. Ground electrical currents associated with the geomagnetic storm caused disruption of telegraph communication systems in Europe and America.

A brilliant auroral display occurred on 24 October 1870. Another huge display—perhaps the most extensive in relatively recent times—occurred on 4 February 1872. Aurora was again visible to the tropics on this occasion, with sightings being made from Bombay and elsewhere in India. The display was also seen from such locations as Aden, Egypt, Guatemala, and the West Indies. The event accompanied a major geomagnetic storm.

Aurorae in the Twentieth Century

Many outstanding auroral storms occurred during the twentieth century, although, as mentioned earlier, the increasing spread of artificial light pollution somewhat restricted the visibility of these, particularly in the latter half of the century. Major events in the early twentieth century included those of 25 September 1909 (visible from the Cocos Islands, and from Singapore at a latitude of 1°25′N), and 15 May 1921 (seen in Samoa at 14°S).

A well-remembered auroral storm was that of 25 January 1938, seen widely across a Europe on the brink of war. The display was strikingly red, and was easily visible from Cornwall in southwest England. Sightings were also made from Barcelona in Spain and Lisbon in Portugal.

The International Geophysical Year (IGY) in 1957–1958 was timed to coincide with the peak of sunspot cycle 19, which was attended by extremely high auroral activity. Displays on 13 September and 23 September 1957 were visible from Mexico, as was the "Great Red Aurora" of 10 February 1958. This last event was accompanied by electrical blackouts in several areas of northeastern Canada, as a result of associated ground-induced currents.

The maximum of the following cycle 20 was relatively quiet, though a notable series of events did occur in August 1972, again accompanied by geomagnetic effects that caused power fluctuations and communications difficulties in North America. It is fortunate that the violent solar activity responsible for setting off these events did not coincide with any of the Apollo lunar missions: the energetic particles released would have been lethal to astronauts aboard their relatively unshielded vehicle. The effects of this outburst of activity on the interplanetary medium were measured from a number of unmanned spacecraft, including the Pioneers (then en route to Jupiter) and by radio astronomers observing distant sources through the disturbed outer solar atmosphere.

Cycle 21 brought several good displays of aurora to mid-latitudes, particularly during 1978–1979, and again in 1982. Noteworthy events for observers at lower latitudes included major storms in May and November 1978, and an all-night coronal storm with red rays visible right down to the south of England on 1–2 March 1982. Ironically, the most extensive and spectacular display of this entire sunspot cycle occurred almost at its minimum, on the night of 8–9 February 1986, visible from Hawaii.

The Great Aurora of 13–14 March 1989

Sunspot cycle 22, which peaked in June 1989, was marked by a very rapid rise in sunspot activity, accompanied by several fine aurorae for observers at lower latitudes.

Of the aurorae during this period, the most dramatic was undoubtedly that seen widely over 13–14 March 1989. The display was sufficiently bright to be visible even from quite badly light-polluted locations (some observers at dark sites claimed that the aurora cast shadows!), and attracted considerable public attention.

The source of this storm was a large, complex flare-active sunspot group in the Sun's northern hemisphere. The spot was readily visible to the protected naked eye for several days either side of the auroral storm itself. Listed by NOAA as Active Region (AR) 5395, the spot group first appeared round the eastern limb of the Sun on 7 March: flare activity extending over the limb from the far side of the Sun had already alerted solar astronomers to the possibility that this was an exceptional active area. Several further flares—many of them very violent—were seen in association with the spot group, including one on 10 March, which appears to have been associated with the ejection of energetic particles whose arrival at Earth caused the auroral storm.

The aurora, following the observed flare by 36–48 hours, was extensively seen. Initial sightings of auroral activity, coincident with the onset of worldwide magnetic field disturbances, were made from County Clare, Ireland, in the early hours (about 02 hours Universal Time) of 13 March. Activity built through the day, providing a great spectacle for the many around the world who witnessed it.

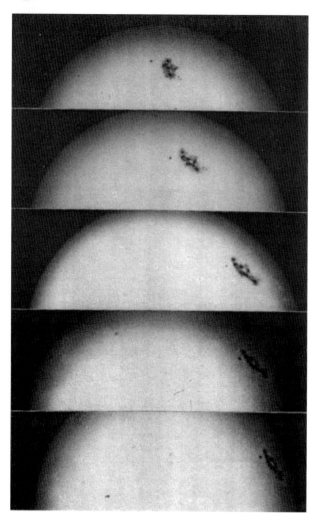

Figure 5.1. A series of images by the late Bruce Hardie showing the development of giant sunspot group AR 5395 as it was carried across the visible disk by the Sun's rotation in March 1989. Solar flare/coronal mass ejection activity associated with AR 5395 triggered the Great Aurora of 13–14 March 1989, seen extensively around the world.

Observers in Florida, Arizona, Southern California, and elsewhere in the southern United States, were surprised to see blue, green and red auroral forms filling their skies. For many, the most memorable feature was the appearance of vivid red auroral rays: as is evident in records stretching back to medieval times, red coloration is characteristic of the most extensive aurorae. Even Canadian observers, many of them used to regular, fine auroral activity, found this event remarkable.

Sightings were also made from Mexico and from the Caribbean. The aurora was recorded from Grand Cayman Island at latitude 19°N.

In the antipodes, the Aurora Australis was visible from New Zealand and Queensland, Australia (latitude 25°S). The display was also seen from South Africa, at latitudes of 30°S, where the aurora is a rare sight indeed.

European observers also witnessed a fine display—reputedly the best in the Netherlands since the 1938 storm. In the British Isles, skies were clear over most of the country for once, and the aurora was visible from points between the Orkneys in the north and the Channel Islands in the south. Hundreds of detailed reports from amateur astronomers were received by the British Astronomical Association. Many accounts were also received from the general public, the following by a Mrs. Wylde of Malmsbury, Wiltshire, being typical of how the display was seen by the many for whom this was their first aurora:

> "The far eastern end of the display began to develop a very strong glow. This bright blob stabilised, no longer pulsating, but growing in size. Suddenly it flashed out several white rays, quite bright and reaching towards the zenith. Then came the Colour—all the rays blushed a bright crimson, and for a second or two I saw a rich blue and a trace of glorious golden yellow. The rays, ranging from bright crimson to deepest dull red, came right over the zenith, and seemed to converge. The display grew from horizon to horizon."

This coronal phase, usually short-lived from mid-latitudes, persisted for much of the night for observers in Scotland, for whom the display continued all night, before fading into the dawn. The extreme peak of activity came just before magnetic midnight, around 22 hours local time, and was the phase witnessed by the majority of British observers. At the same time, the aurora was visible to observers in Hungary, Portugal, and Spain, and into the Mediterranean area.

Most British national newspapers reported on the aurora. For example, *The Times* covered it under the headline "Observing the light fantastic." Elsewhere, *The Guardian* described the aurora as a "Spring highlight for the south." Local newspapers also covered the aurora. The Cambridge *Evening News* spoke of a "Spectacular sight in south for sky searchers." Aurorae are relatively uncommon in the most densely populated, southernmost, parts of the United Kingdom, and it is not surprising that this auroral display was such a source of media interest.

In addition to its visual splendor, this auroral event was accompanied by intense radio effects as the particle influx disrupted the ionosphere high in the atmosphere. Amateur radio operators enjoyed record-breaking communications over long distances— England to Italy, for example, impossible under normal conditions using simple equipment.

The geomagnetic disturbance on 13–14 March was also intense. The ground-induced currents associated with this activity caused electrical blackouts in Sweden and Canada, as a result of power surges in high-latitude areas. In Quebec Province, 6 million people were left without electricity for up to nine hours—testimony, indeed, that geomagnetic storms can have important effects on human affairs.

Visual observers at higher latitudes saw activity continue into the next night, though at gradually diminishing levels. Within a couple of days of the storm, all was quiet. Some waited with great anticipation for the return of Active Region 5395 one solar rotation later, and the possibility of a recurrence of the intense aurora in early April 1989. By then, however, the sunspot had decayed, and nothing out of the ordinary was seen in terms of aurora.

Other Events in Cycle 22

As discussed in Chapter 3, there appear to be two peaks of auroral activity at lower latitudes in most sunspot cycles. The Great Storm of 13–14 March 1989 was, without doubt, the most spectacular event in the first peak associated with sunspot cycle 22.

The display had been preceded, for those at higher latitudes at least, by good auroral activity throughout the autumn of 1988.

Following the Great Storm by a matter of weeks was a further extensive display, also visible to the latitudes of southern England, on 25–26 April 1989. Thereafter, auroral activity at lower latitudes dwindled for a time: with the exception of a faint aurora on 28–29 July, briefly visible to the latitudes of southern England, 1990 was a very quiet year for aurorae at lower latitudes, despite continued high sunspot numbers.

This comparative lull was brought to an end by a vigorous auroral display, seen extensively in the United States on 23–24 March 1991, and continuing into the following night (24–25 March), when it was visible right down to the South Coast of England. From Scottish latitudes, this aurora, with interludes of brilliant red coloration, was coronal for much of the night. The display was triggered by flare activity associated with the long-lived Active Region 6555, the second-largest [covering 3000 millionths of a solar hemisphere (MSH)] spot group of cycle 22, which had crossed the Sun's central meridian a few days previously.

Another extensive (2500 MSH) spot group, AR 6659, crossed the Sun's disk on 9 June 1991, and was reportedly even more flare-active than that responsible for the Great Aurora in 1989 March. Forecasts, widely picked up on by the news media, suggested that aurora would be visible to very low latitudes as a consequence. In the end, the anticipated events around 12 June amounted to comparatively little, except for those in the United States who had darker skies than their colleagues at the latitudes of the British Isles. Activity was extensive for only a short time.

Figure 5.2. The major aurora of 24–25 March 1991, photographed over Chichester (the City's Cathedral spire is visible at lower right), West Sussex in southern England. Image: Neil Bone

The expectation of good auroral activity as 1991 drew to a close was borne out by an event visible to northern Italy on 1–2 October, and the exceptional 8–9 November display. Late December of 1991 was also graced by some fine, if less extensive, events, the outbursts of activity ending with another short-lived display, this time over western European longitudes, on 1–2 February 1992.

The Major Aurora of 8–9 November 1991

That the aurora can sometimes spring surprises is perhaps best demonstrated by the outstanding event of 8–9 November 1991. Activity during the preceding months had admittedly been high, with several fine displays extending to lower latitudes. During the first week of November, however, the Sun appeared relatively quiet, with the white-light projected image showing only a few, relatively small, spot groups. Despite this comparative dearth of sunspots, it became apparent, as soon as darkness fell, that a major aurora was in progress. Like the Great Aurora of March 1989, this event was witnessed by many thousands of amateur astronomers and others around the world.

Intense auroral activity was visible from early evening onwards, even at the latitudes of the English Midlands. A group of beginners, gathered for an astronomy course at the Preston Montford Field Centre near Shrewsbury, was treated to spectacular views of brilliant red auroral rays filling the northern sky. As the night went on, the aurora intensified, and even along the South Coast of England, the aurora stretched to the zenith at times. The ray-tops were visible from the south of France.

In southwest Scotland, experienced observer Tom McEwan recorded coronal activity lasting all night, with an auroral band running through Orion's Belt to the *south* of the sky. Quite simply, this was a stunning display, perhaps even more intensely colored than the March 1989 storm, and a good target for photographers. A great many pictures were taken by amateur astronomers, showing the dominant red rays and green bands.

Activity continued unabated as night fell over the United States some hours later. The aurora was visible down to the southern states of Texas, Georgia, Alabama, and Oklahoma.

The major aurora of 8–9 November 1991 came at a particularly poignant time for magnetospheric physicists at the University of Iowa, four of whose colleagues (Chris Goertz, Dwight Nicholson, Bob Smith, and Yinhua Shan) were shot dead the previous week by a disaffected graduate student. Memorial services for the four were held close to the date of the aurora; scientists at Iowa regard the aurora as having been an appropriate natural tribute in itself.

No obvious candidate sunspot group appeared to be in the right place on the solar disk to set off this major auroral storm. Rather, it is believed that a disturbance—perhaps a flare in a spot group on the Sun's averted hemisphere—produced a shock wave, which in turn caused detachment of a filament (equivalent to a solar prominence seen in profile against the bright disk). A candidate filament disappeared from the Sun's SE quadrant on 6 November. Material from the filament, thrown into the solar wind, arrived at Earth around 48 hours later, resulting in the magnetospheric disturbance and spectacular auroral display recorded on 8–9 November. Pre-dating the continuous

Figure 5.3. Strong aurora on 8–9 November 1991, with long red-topped rays stretching to Polaris even from the relatively low latitudes of Sussex. Image: Neil Bone.

monitoring by spacecraft such as SOHO, this event "blind-sided" astronomers in a manner unlikely to be repeated in the future!

The Strange Case of Cycle 23

Following the decline of solar and geomagnetic activity up to sunspot minimum in 1996, the start of new cycle 23 was anticipated with interest. Sunspot numbers began to increase significantly during 1999, and it was expected that if the pattern seen in cycles 21 and 22 was followed, some strong auroral displays penetrating to lower latitudes would soon occur. In fact, it took until April 2000 for major activity to show. Also, unlike the situation in cycle 22 when the Great Aurora of 13–14 March 1989 was preceded by a marked build-up in the number of displays reaching, say, the latitudes of central Scotland, there was very little to suggest that auroral activity was taking off at all.

Figure 5.4.
Strong red and green oxygen emissions made the 6–7 April 2000 aurora a spectacular sight across the British Isles. Image: Neil Bone.

The Aurora of 6–7 April 2000

The drought in low-latitude auroral displays was finally broken four years after the start of cycle 23. A halo CME was recorded by the SOHO spacecraft on 4 April 2000, associated with flare activity in a sunspot group designated AR 8933 at 1541 UT on that day. Starting at 17h UT on 6 April—a little more than 48 hours later—the CME impacted on Earth's magnetosphere, triggering a major geomagnetic storm. An hour earlier, increased particle flux and gusting in the solar wind were recorded by the ACE spacecraft, as was the onset of a sustained period of strong southerly directed Interplanetary Magnetic Field—perfect conditions for bringing the aurora to lower latitudes.

Observers in eastern Europe were first to see the auroral display, which sent long red rays high into the spring sky. Sightings were made from Estonia, southern Finland, Germany, and the Czech Republic. By nightfall over western Europe, around 2100 UT, the storm was in full fling, and hundreds of amateur astronomers—and

Figure 5.5. Red auroral light suffusing the southern sky from southern England during the 6–7 April 2000 aurora; the display was visible down to the latitudes of Portugal.

other more casual skywatchers—from Scotland to the south coast of England enjoyed the spectacle. Activity ebbed and flowed, with outbursts of strong activity interspersed by relative lulls. The display was bright enough to be visible even from some quite badly light-polluted locations—York city center, for example. A common theme in observers' reports was mention of bright green "curtain" bands, sending up long red rays at the most active periods. At higher latitudes—central Scotland and the Isle of Man—the aurora remained in coronal form for several hours around the peak of activity, which came between 2300 and 2320 UT. Thereafter, activity gradually declined, but the display was still strong as night fell over the eastern seaboard of North America: the aurora was well seen from New York and New Jersey.

At its peak, the aurora suffused the entire sky with a reddish glow, even from the south of England. The display was visible as far south as Portugal, and was also extensively seen in Ireland, Belgium, Holland, and Austria.

The Bastille Day Event, 15–16 July 2000

Following the April 2000 aurora, the Sun became increasingly active and Space Weather forecasters highlighted several potential storms through June, none of which, sadly, resulted in widely seen auroral displays. The most significant event of 2000 came in July. On 14 July (Bastille Day) at 1024 UT, a major (class X5.7) solar flare occurred in the large, magnetically complex spotgroup AR 9077. Soon afterwards, SOHO's coronagraphs detected a full-halo CME heading straight toward Earth; AR 9077, close to the central meridian of the visible solar disk, was in a highly geoeffective position, and the CME shockwave, traveling extremely rapidly (1300–1800 km/s) arrived in near-Earth space within 30 hours of the observed X-ray flare. The event triggered a strong solar radiation storm, and the CME's arrival resulted in an intense —"Extreme" on NOAA's five-point scale—geomagnetic storm during which detectors aboard the ACE spacecraft were saturated. A casualty of the geomagnetic storm conditions was the Japanese Advanced Satellite for Cosmology and Astrophysics (ASCA), sent tumbling out of control—thankfully having successfully completed its mission.

For observers on the ground, the aurora was at times spectacular, despite summer twilight conditions at high northern latitudes, and glare from a Full Moon. In the British Isles, conditions were rather cloudy for many, but the aurora's strong red and purple rays were seen even as far south as Sussex and Kent. Observers in the southern United States, including Texas, Georgia, and Florida saw extensive displays of strongly colored aurora.

March–April 2001—A Rash of Sunspots

Sunspot cycle 23 showed a major outburst of activity in the opening quarter of 2001, including in the third week of March, the giant complex AR 9393. Growing to cover 2400 MSH by the time it was crossing the central meridian of the Sun's visible disk around 28–29 March this spotgroup was the largest seen since 1991. AR 9393 was the site of several X-class flares, and the SOHO coronagraphs detected halo CMEs associated with such events on 28–29 March. X-class flare activity on the latter date produced a solar radiation storm, and the arrival—so close together as to be a combined hit—of the CMEs at Earth early on 31 March initiated a major geomagnetic storm. The shockwave swept past the ACE spacecraft at 0030 UT, and the disturbance continued for the next 24 hours, with aurora visible as far south as Mexico, and extensive sightings also being made from California, Texas, Arizona, and North Carolina. Observers in New Zealand had a good view of the activity, as—12 hours later—did those in Germany and France. Weather conditions over the British Isles were poor, but the aurora was seen from parts of central Scotland. At all locations, this major display was characterized by intense red and green oxygen emissions.

AR 9393 remained flare-active as the Sun's rotation carried it away from a geoeffective position. An X20 flare on 2 April at 2150 UT was, up to that time, the most powerful yet recorded. Despite occurring close to the Sun's western limb, the flare still produced a strong solar radiation storm, and the associated CME struck Earth's magnetosphere a glancing blow—insufficient to set off major auroral activity—a couple of days later.

As AR 9393 departed, another flare-active group rotated into view. AR 9415 was, during the first week of April 2001, a source of more M- and X-class flares one of which caused a solar radiation storm on 10 April. A CME associated with an M8 flare in AR 9415 on 9 April at 1535 UT swept into near-Earth space on 11–12 April, setting off a severe geomagnetic storm which lasted for 22 hours. Observers across much of Europe had very fine views of green and red aurora, with purple nitrogen emissions at

Figure 5.6. The aurora of 21–22 October 2001 showed pronounced red colouration at the tops of rays. This image, taken from Sussex in southern England, shows strong green (557.7 nm) oxygen emission in the lower parts of the aurora, through which the familiar stars of the Plough are visible. Image: Neil Bone.

the tops of the long rays. The aurora was overhead and coronal from central Scotland. Southern England was largely cloudy, but the display was seen and photographed from Cornwall and Sussex. The aurora was described as the best of many years for observers in Germany, France (where it was visible from the Mediterranean coast), Denmark, and the Netherlands. Later stages of the activity were visible from California and New Mexico.

AR 9415 was the source of yet another large—X14—flare on 15 April, when close to the Sun's western limb. Meanwhile, having survived passage around the averted hemisphere, AR 9393—re-designated by NOAA as AR 9433—returned at the eastern solar limb on 19 April, somewhat diminished in extent but still active. A CME from the spotgroup produced a minor geomagnetic storm on 28 April.

After this 4–5-week bout of occasionally intense activity, the next several months were relatively quiet geomagnetically. X-class flare activity associated with AR 9661 on 19 October brought about one further well-reported auroral display at lower latitudes on the night of 21–22 October 2001. Visible to the south of England, and from Spain, France, and Germany, this aurora was notable for interludes of vivid red rayed activity. Observers at lower latitudes then saw little more in the way of aurora for a while, with no particularly outstanding events through 2002 or the first half of 2003. A major surprise awaited, however, as events unfolded in October and November of 2003.

The Hallowe'en Storms of October 2003

The interval from late 2001 until October 2003 was far from devoid of sunspot activity. Comparatively few major solar flares or CMEs occurred, however, and most of the aurorae seen were visible from higher latitudes. Gradually declining sunspot activity made it easier to note the effect of coronal hole streams—again, a source mainly of higher-latitude aurorae. From the middle of 2003, cycle 23 began to produce numbers extensive of active regions. Surely most remarkable among these were the three giant spotgroups which brought October 2003 to a spectacular close.

First indications of the coming disturbance were noted from the SOHO spacecraft, which detected a series of CMEs emerging from beyond the Sun's eastern limb. The source of these was AR 484, which rotated onto the visible hemisphere on 18 October, This complex spotgroup reached its maximum extent—1750 MSH—on 22–23 October, and was the seat of numerous M-class flares, and an X-class flare on 19 October.

By 22 October AR 484 had been joined on the Sun's visible hemisphere by the even larger, and more active, AR 486. AR 486 reached a maximum extent of 2600 MSH (slightly bigger than March 1991's AR 9393) on 29–30 October, and was the source of seven X-class flares. A third giant spotgroup, AR 488, emerged on 27 October. During the final week of October, the unusual situation obtained of three sunspots visible to the protected naked eye!

Between them, these produced a total of 44 M-class flares and 11 X-class flares in a 17-day interval. Normally, this late in the sunspot cycle, one might expect fewer than a dozen X-class flares in an entire year! Most remarkably, perhaps, a 26-hour period on 2–3 November saw the occurrence of three X-class flares.

In terms of flares, AR 486 proved the most active of cycle 23. A huge X17.2 flare erupted there on 28 October at 1111 UT, triggering a major geomagnetic storm and auroral displays over Australia and the southern United States on 29 October. The disturbance continued into the night of 29–30 October when all-sky aurora was extensively visible from the northern half of the British Isles.

Figure 5.7.
October 2003's giant sunspot groups AR 484, AR 486 and AR 488. Solar flare/CME activity associated with these triggered the most violent geomagnetic storms of sunspot cycle 23. Image: Ken Kennedy.

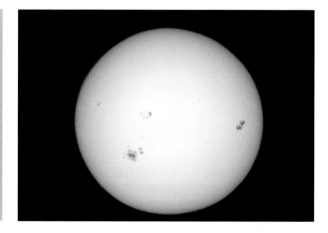

Figure 5.8. Spectacular coronal aurora over central Scotland on 29–30 October 2003. The figure in the foreground is a statue of Sir David Stirling, on a hillside near Falkirk (a popular observing spot for members of the Association of Falkirk Astronomers). Image: Russell Cockman.

Figure 5.9. Another view of the 29–30 October 2003 aurora from near Falkirk, showing extensive rayed bands spreading across the southeastern sky (where Orion is rising). The horizon appears distorted in this semi-fisheye image, taken by Russell Cockman.

Figure 5.10. Auroral rayed band over Sussex late on 29–30 October 2003. Image: Neil Bone.

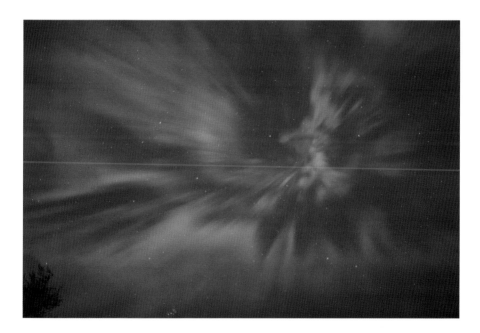

Figure 5.11. Coronal aurora during the 29–30 October 2003 'Hallowe'en Storm', recorded from Aberdeenshire, Scotland. Image: Phil Hart.

Geomagnetic activity was maintained at a high level by a further, X10 flare at 2049 UT on 29 October. Successive nights of extensive lower-latitude aurora occurred on 29–30 and 30–31 October, and—by virtue of their proximity to the date—these have come to be known as the 'Halloween Storms'. The 29 October geomagnetic storm was rated "Extreme," the highest level on NOAA's five-point scale, and an intensity which might only be expected once in an 11-year sunspot cycle.

Among the effects associated with the geomagnetic storm, ground-induced currents led to a power outage affecting 20,000 homes in Malmo, Sweden. GOES satellite measurements of energetic protons in near-Earth space showed a strong solar radiation storm persisting from 22–23 October, peaking 26–28 October, finally declining to normal, quiet levels around 12 November. On a couple of occasions on 29 and

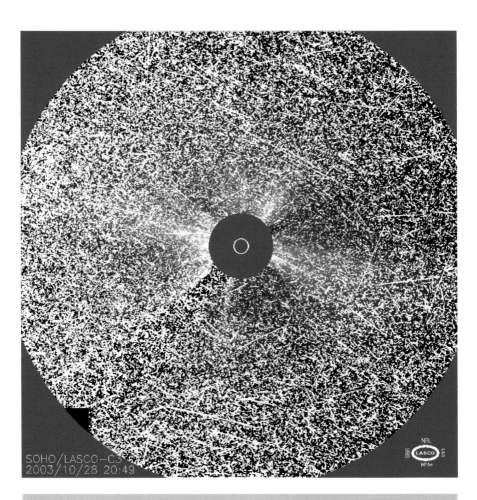

Figure 5.12. Energetic protons ejected during the Halllowe'en Storm events temporarily blinded SOHO's LASCO C3 coronagraph in October 2003. Image: ESA.

30 October, solar cosmic ray "hits" virtually saturated the C3 coronagraph detector aboard SOHO.

The effects of this intense burst of stormy space weather were also felt farther out in the Solar System. En route to its highly successful mapping mission at the Red Planet, the Mars Express spacecraft had its navigational star tracker temporarily blinded by proton hits. Meanwhile, the Cassini probe approaching Saturn detected the radio noise from the 28 October and 4 November flares. Out on the edge of the heliosphere, the Voyager 1 spacecraft also felt the influence of the shockwaves propagating through the solar wind. An increase in the flux of MeV-energy ions (of solar origin) and a decrease in that of—extrasolar—Galactic cosmic rays was found between 14 August and 19 November 2004. At a distance of 90 AU from the Sun, and some 10–12 months after their occurrence, the influence of the Hallowe'en Storm flares was still being felt!

As it departed the western solar limb, AR 486 really let go! On 4 November from 1929 to 1950 UT, a flare estimated to have reached X28 intensity—the biggest ever recorded—erupted. Despite striking Earth's magnetosphere with only a glancing, sideways blow, the associated CME still set off a significant geomagnetic disturbance: had it been directed fully Earthwards, the effects would surely have been extremely dramatic.

In terms of geomagnetic activity, the Hallowe'en Storms of 2003 appear to have outstripped anything previously recorded. The peak three-hourly aa index reached 715 early on 29 October 2003, higher than the 680 attained during the 1921 May storm, and the 698 reached in the major events of 1958 July and 1972 August.

Spacecraft measurements showed that the Hallowe'en Storms caused a major distortion of the Van Allen Belts and Earth's plasmasphere. The outer Van Allen Belt was pushed inwards to about 10,000 km above Earth's surface (its center is usually at an

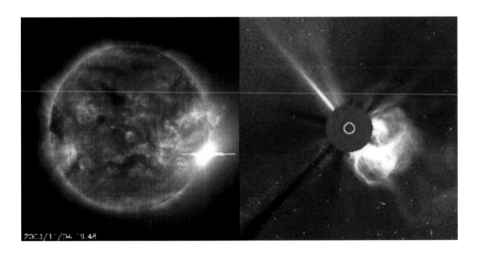

Figure 5.13. SOHO EIT image (left) of the 4 November 2003 X28 solar flare with—a few hours later—a LASCO C3 coronagraph image (right) of the associated coronal mass ejection. Image: ESA.

Table 5.1. The Biggest Geomagnetic Storms

Date	Peak 3-hour aa Index	24-hour Aa Index
24–25 October 1870	464	189
4 February 1872	658	230
25 September 1909	658	329
14–15 May 1921	680	257
22 January 1938	656	240
25 January 1938	656	188
28 March 1946	656	322
22 September 1946	656	272
21 January 1957	490	130
10–11 February 1958	568	298
8 July 1958	698	305
4–5 August 1972	698	215
8–9 February 1986	578	244
13–14 March 1989	365	348
15 July 2000	440	207
29 October 2003	715	298
20 November 2003	578	228

Data from http://www.geomag.bgs.ac.uk/gifs/aaindex.html

altitude of about 25,000 km), and the "slot" region between the inner and outer belts—normally "empty"—became heavily populated with high-energy particles, as detected by the Solar Anomalous and Magnetospheric Particle Explorer (SAMPEX) satellite. This distortion and disturbance persisted for more than a year after the storms. The IMAGE (Imager for Magnetosphere-to-Aurora Global Exploration) spacecraft, meanwhile, is able from high orbit to visualize the plasmasphere via extreme ultraviolet emissions from resident ionized helium (He^+) at 30.4 nm wavelength. Observations showed the plasmasphere's outer boundary to have been pushed in from 19,000 km to as close as 6000 km above Earth's surface around 31 October. Unlike that affecting the Van Allen Belts, this distortion of the plasmasphere was relatively short-lived, lasting only a few days. The overall effect was to weaken the magnetospheric "shield" against incoming energetic particles, making the near-Earth space environment a hostile place for satellite operations. The increased flux of solar cosmic rays was also treated as a hazard by commercial aircraft operators for flights at altitudes in excess of 7500 meters.

Among other effects, depletion of stratospheric ozone at high latitudes in the northern hemisphere during the winter of 2003–2004 has been partially attributed to the violent solar activity of 2003 October. A study by a team from the Laboratory for Atmospheric and Space Physics at the University of Colorado, Boulder, found that ozone concentrations at altitudes of 40 km in the Arctic were at about 40% of their normal springtime value in early 2004. This depletion appears to have come about through a combination of factors. Accelerated electrons entering the high polar atmosphere during the Hallowe'en Storms collided with and ionized nitrogen which in turn reacted with oxygen to produce nitrogen oxide and nitrogen dioxide. Satellite

Figure 5.14. Coronal aurora over Falkirk, central Scotland on 20–21 November 2003. Image: Malcolm Gibb.

observations in November and December 2003 showed an increase in the concentration of these "Nox" species. Carried downwards by the polar vortex circulation of the high atmosphere, Nox molecules were able to react with, and destroy, stratospheric ozone. The effect was restricted to the (dark) Arctic stratosphere; in the Antarctic, sunlight destroyed Nox molecules before they could be transported to the stratosphere. It is also likely that fast solar protons had a role in the observed ozone depletion.

The three giant spotgroups survived long enough to return, still active, to the Sun's visible hemisphere from the second week of 2003 November. Flare activity from AR 484 sent further CMEs Earthwards, triggering a major geomagnetic storm starting early on 20 November. Auroral activity was visible to Florida, Alabama and New Mexico. Later, observers in Scotland, Northern Ireland and the north of England also witnessed a fine display. Indeed, the aurora on 20–21 November 2003 was visible as far south as Italy and—unusually, for the second time in a matter of a few weeks!—Greece.

As they followed AR 484 onto the solar disk, AR 486 and AR 488 began to decay and diminish in activity, finally bringing to a close the most vigorous interval of geomagnetic disturbances in sunspot cycle 23.

A Sting in the Tail

Cycle 23 was, however, still slow to decline. Observers continued to record numerous sunspot groups/active areas on the projected solar disk throughout 2004. In January 2005, AR (10)720 brought another revival in flare activity, producing several significant X-class events. Reaching a maximum size of 1880 MSH, AR 720 was the source of CMEs which set off a strong geomagnetic storm on 17 January, bringing bright aurorae to Canadian and northern US skies. A major event pushed the aurora further equatorwards on 20–21 January, when observers in western Europe had excellent views. The associated solar radiation storm, produced by high-energy protons from the Sun, was the strongest since October 1989.

With sunspot minimum expected a year or so hence, solar activity in 2005 continued to surprise. May brought another big spotgroup, AR (10)759, source of several major flares and CMEs triggering an extreme geomagnetic storm on 13–14 May, when aurora was visible as far south as California. Growing as the Sun's rotation carried it beyond the limb, AR (10)786 in June became strongly flare-active when in a non-geoeffective position. Conversely, AR (10)798 grew rapidly (reaching 930 MSH) as it crossed the Sun's Earth-facing hemisphere, unleashing nine X-class flares up to 7 September, leading to enhanced high-latitude auroral activity. On its return in early October, AR 798 had decayed leaving only a patch of faculae.

One last blast of activity as cycle 23 finally began to wind down came with AR (10)822, which decayed as it crossed the visible solar disk in 2005 November.

Thereafter, only minor spots were in evidence as the cycle sank toward the anticipated mid-2006 minimum. Indeed, runs of up to ten days at a time with no apparent sunspot activity whatsoever became the hallmark of the opening months of the year, and a corresponding lack of low-latitude aurorae ensued. What little auroral activity was seen resulted—as is typical for that phase in the solar cycle—from high-speed coronal hole streams. Observers at higher latitudes recorded quiescent glows and arcs, and there were none of the fast-moving, rapidly changing "curtain" displays which so thrilled thousands of witnesses a couple of years earlier.

The Next Two Cycles

Using a combination of modern technology and past observational records, solar scientists have tried to forecast how future sunspot cycles may perform. Helioseismological studies from SOHO and other sources show that at a depth of about 200,000 km in the Sun's convective zone there is a roughly 40-year "conveyor belt" circulation of plasma in either hemisphere, cycling between low and high latitudes. Magnetic flux emerging from this circulation and breaking through the photosphere, as described in Chapter 2, is the source of sunspot activity.

The speed of the conveyor belt is variable, and can be inferred by studying the rate at which the zones of sunspot activity migrate from higher solar latitudes (early in the cycle) toward the equator (late cycle). Rapid circulation results in greater magnetic flux transport and higher numbers of sunspots. Use of sunspot data stretching back over a century suggests that the rate of zonal migration can be used to predict the intensity of activity two cycles hence.

On this basis, using sunspot observations from 1986 to 1996, a team led by Mansumi Dikpati at the US National Center for Atmospheric Research has predicted that sunspot cycle 24, peaking in 2010–2011, should be intense, and at least 30% more active than cycle 23. At peak, it may rival the strongest sunspot maximum so far observed, that of cycle 19 in 1957–1958.

Conversely, NASA solar physicist David Hathaway points to cycle 25 as potentially one of the weakest ever, based on sunspot migration observations from cycle 23. It would appear that the conveyor belt circulation slowed considerably by the early 2000s.

This is seen as good news for projected manned Mars missions that may take place around the time that cycle 25 is nearing peak in 2022. There will then be less risk from intense solar radiation storms. On the other hand, weak sunspot activity and a less vigorous solar wind might lead to increased Galactic cosmic ray flux in the heliosphere.

The long-range Space Weather forecast, then, promises much for aurora observers in cycle 24, but leaner times in cycle 25! If cycle 24 has the rapid take-off in sunspot activity seen in past very active cycles, it is likely that 2008-09 could see a strong "early" peak in low-latitude auroral occurrence.

CHAPTER SIX

Aurora Elsewhere

Viewed from afar, Earth is a strong emitter of radio waves in the range from below 100 kHz to nearly 1 MHz, peaking around 300 kHz. This naturally occurring Auroral Kilometric Radiation (AKR) is stronger than any man-made transmissions at such frequencies, averaging 10 million watts and sometimes peaking at a billion watts. The AKR is not, however, detectable at ground level, being reflected by the ionosphere's E-layer at 110 km altitude. Several of the other major planets of the Solar System also have their own auroral signatures. Observation of these is beyond amateur means, but it is of interest to compare the aurorae that occur at, say, Jupiter and Saturn, with that of our homeworld.

Spacecraft have explored the near environments of all the major planets with the exception of Pluto (which many astronomers contend is not a major planet in any case!). This exploration has revealed all to possess magnetic fields of differing intensities, which interact with the solar wind to varying extents. Studies of these offer important insights to the behavior of the terrestrial magnetosphere.

Particles accelerated in their magnetospheres give rise to auroral displays in the atmospheres of the gas giants Jupiter, Saturn, Uranus, and Neptune. These displays, on a vast scale, have been imaged on the planets' night-sides by the passing Voyager spacecraft.

Jupiter

Jupiter has by far the most extensive magnetosphere in the Solar System, with a bow shock lying 6 million km (over 80 Jupiter-radii) upwind, and a magnetotail extending downwind beyond the orbit of Saturn. From the distance of Earth (4 AU at opposition),

Jupiter's magnetosphere subtends an angle of about 1° (twice the apparent diameter of the Moon) against the sky surrounding the planet. The precise dimensions of the jovian magnetosphere are highly variable, ranging from 50 to 100 Jupiter-radii (3.6 to 7.2 million km) depending on solar wind intensity.

Trapped plasma in Jupiter's magnetosphere is concentrated in an equatorial plane *magnetodisk* as a result of the system's rapid rotation (slightly less than 10 hours). Measurements obtained during the passage of the Ulysses spacecraft through near-Jupiter space in 1992 February indicate that the jovian radiation belts extend to a maximum latitude of 40° (compared with 70° for the terrestrial Van Allen belts). The jovian magnetospheric plasma population includes electrons from the planet's upper atmosphere and sulfur and oxygen ions (from sulfur dioxide) ejected during volcanic eruptions from the innermost Galilean satellite Io. The latter material forms an equatorial ring, the *Io torus*. Ulysses measurements showed the Io torus to be inhomogeneous, containing "hot spots" of radio emission. The principal role of the solar wind appears to be in shaping, rather than populating, the jovian magnetosphere.

Particles in Jupiter's magnetosphere, by virtue of its immense size, can be accelerated to extremely high energies. Bursts of radio noise resulting from particle motions in the jovian magnetosphere are sufficiently powerful to be detected from Earth. These energetic particles present a hazard for spacecraft in the jovian system.

The Voyagers imaged aurorae on Jupiter's night-side. Particles accelerated into Jupiter's atmosphere at high latitudes give rise to aurorae in ovals girdling either magnetic pole, as on Earth. Jupiter's magnetic axis of offset by 10–15° from the planet's axis of rotation. The jovian aurorae are 1000 times as powerful as their terrestrial counterparts and appear predominantly red thanks to the abundance of hydrogen in the planet's atmosphere.

Observations of Jupiter's aurorae have been obtained on a routine basis using ultraviolet detectors aboard the International Ultraviolet Explorer satellite, and ground-based infrared telescopes. Emissions of the H_3^+ ion, at a wavelength of 3.53 microns in the infrared have been of particular use in such studies. Around this wavelength, Jupiter appears dark as a result of atmospheric methane absorption. Jovian H_3^+ emissions can vary in intensity over timescales of an hour or so.

Together with other phenomena—principally the prominent large dark spots resulting from the impacts—the jovian aurorae were subject to intense scrutiny during the week of 16–23 July 1994, as the fragments of the disrupted Comet Shoemaker-Levy 9 crashed into the giant planet's atmosphere around 44°S. Around the time of the impacts, 20-cm-wavelength synchrotron radiation from electrons accelerated in the jovian magnetosphere was elevated by 25–50%, and remained at abnormally high levels for some months afterwards. Immediate effects related to the impacts included fading of sections of Jupiter's southern auroral oval, and a brightening of the northern oval roughly 45 minutes following the impact of fragment K (Figure 6.2). The latter effect has been attributed to the arrival, along the connecting magnetic field line, of material from the equivalent (conjugate) point in the opposite hemisphere.

Observations of Jupiter's aurorae have been obtained using the Hubble Space Telescope's Wide Field and Planetary Camera (WFPC2) and Space Telescope Imaging Spectroscope (STIS). These have shown the auroral ovals around either jovian

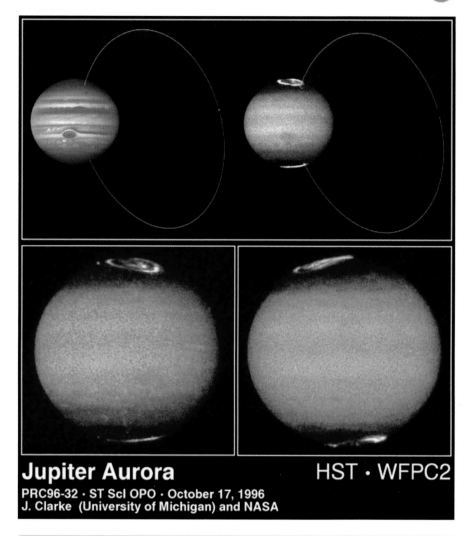

Jupiter Aurora   HST · WFPC2
PRC96-32 · ST ScI OPO · October 17, 1996
J. Clarke (University of Michigan) and NASA

Figure 6.1. Rapid changes in Jupiter's auroral ovals, imaged on 17 October 1996 with the Hubble Space Telescope. The upper panel shows a magnetic field line connecting conjugate points in either hemisphere. Image: J. Clarke and G. Ballater (University of Michigan), J. Trauger and R. Evans (Jet Propulsion Laboratotu) and NASA.

magnetic pole to be mirror images of each other. Connections between the auroral ovals and the Galilean satellites Io, Europa, and Ganymede have been imaged as bright spots—"footprints." Io's footprint is 1000–2000 km across. In contrast with the evening substorms in Earth's auroral ovals, under disturbed conditions Jupiter's ovals undergo their major brightening and broadening in the dawn sector.

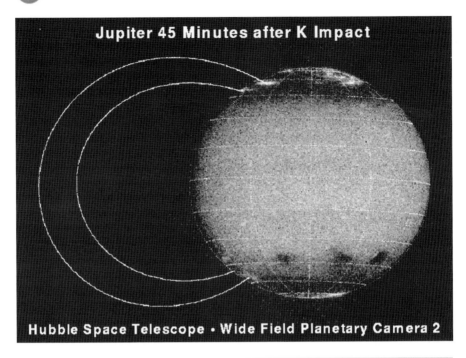

Figure 6.2. In addition to producing prominent dark 'scars' in the planet's cloud-tops, the impacts of the Comet Shoemaker-Levy 9 fragments disrupted Jupiter's magnetosphere and auroral ovals. Following thr impact of Fragment K on 19 July 1994, brightening was observed at conjugate points in he jovian auroral ovals, using the Hubble Space Telescope. Image: Hubble Space Telescope Jupiter Imaging Team.

Saturn

Saturn's magnetosphere was first detected and measured by Pioneer 11 in 1979, and later examined in more detail by the Voyagers. As with Jupiter, Saturn has an extensive magnetosphere, on whose outer fringes orbits the large satellite Titan. Titan's atmosphere is a source of nitrogen which contributes to the particle population of the saturnian magnetosphere. The inner satellites may also contribute material: cryovolcanism on Enceladus is a source of water vapor, for example.

Saturn's magnetosphere contains a partial magnetodisk, and in some respects shows a structure intermediate between those of the Earth and Jupiter. As might be expected, the ring system has a role to play, in absorbing energetic particles: trapped, energetic particles are absent from the region of the magnetosphere bounded by the rings.

Like those of Jupiter, Saturn's auroral ovals have been imaged in the ultraviolet using the Hubble Space Telescope since 1995. Saturn's aurorae extend up to 2000 km above the cloud-tops, and show rapid changes (over a couple of hours) in brightness and extent. As with Jupiter, the main site of activity in the saturnian auroral ovals appears

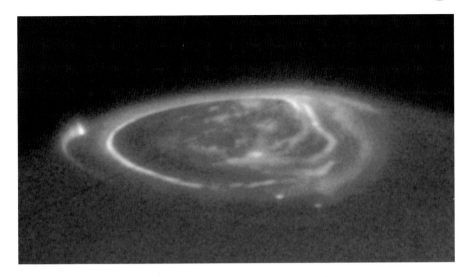

Figure 6.3. An ultraviolet image from the Hubble Space Telescope on 26 November 1998, showing the 'footprints' of the satellites Io, Ganymede and Europa, connected by magnetic flux tubes to Jupiter's auroral oval. Io's footprint is the bright spot at the left-hand limb. Image: NASA/ESA, John Clarke (University of Michigan).

to be in the dawn sector. Interestingly, storm activity causes Saturn's auroral ovals to shrink toward the poles (rather than expanding towards the equator like Earth's ovals). Bright storm activity can last for several days, and during such events Saturn's auroral ovals take on a spiral configuration.

Uranus and Neptune

Uranus is unusual in having an axis of rotation greatly tilted relative to the orbital plane. The magnetic axis of the planet is also considerably offset from its rotational axis. Particle abundances measured by Voyager 2 during its close approach to Uranus in January 1986 indicate the magnetosphere to be populated chiefly by material from the planet's upper atmosphere. As with the saturnian system, the plasma population in Uranus' magnetosphere is controlled to some extent by the planet's satellites and material in the tenuous ring system. Collisions between accelerated particles and the ring material may account for the darkness of the latter. Voyager detected auroral hydrogen emissions in Uranus' upper atmosphere.

In a number of respects, Neptune—which Voyager 2 encountered in August 1989—is quite similar to Uranus. Neptune's axis of rotation is reasonably upright, but again, and surprisingly, its magnetic axis was found to be markedly offset. The neptunian magnetosphere is populated by particles from the planet's atmosphere, and from that of the main satellite Triton. Ionic species include hydrogen and nitrogen. Voyager

Figure 6.4. Saturn's auroral ovals imaged from the HST using the Space Telescope Imaging Spectrograph (STIS) in the ultraviolet, October 1997. Image: J.T. Trauger (Jet Propulsion Laboratory) and NASA.

detected weak auroral activity in Neptune's atmosphere, at an intensity much less than that of terrestrial aurora. Aurora is also generated in the tenuous atmosphere of Triton by particles accelerated in the neptunian magnetosphere. Like those of Uranus, Neptune's thin rings are darkened by particle bombardment.

The offset of their magnetic axes has been taken to imply that the magnetic fields of Uranus and Neptune are generated in layers of those planets' interiors *above* their cores. Much further investigation will undoubtedly be devoted to the mechanisms by which these fields are produced.

The magnetic fields of both Uranus and Neptune are also exceptional in being markedly tilted with respect to the planets' axes of rotation. Uranus' magnetic axis is tilted at 59° with respect to the axis, that of Neptune at 47°. Both values are markedly greater than the tilts for any of the other planets.

Figure 6.5. Ultraviolet images from the STIS aboard the Hubble Space Telescope, showing development of auroral activity on Saturn on 24, 26 and 28 January 2004. Unlike their terrestrial counterparts, Saturn's auroral ovals shrinking towards the poles under disturbed conditions. Image: NASA, ESA, J. Clarke (Boston University) and Z. Levay (STScI).

Mercury

By virtue of its proximity to the Sun, Mercury experiences the most intense solar wind of all the planets. Despite its small size (4880 km diameter), Mercury has a relatively strong magnetic field, suggesting the existence of a proportionally large metallic core. The magnetic field of Mercury produces a bow-shock upwind of the planet. Mercury is large compared to its magnetosphere, and the planet's solid body fills the volume of space where trapping regions equivalent to Earth's Van Allen belts might otherwise form. Neither, in the absence of an appreciable atmosphere, does Mercury possess an ionosphere.

Venus

Venus, in many respects regarded as Earth's twin despite its patently inimical atmosphere and searing temperatures, rotates only slowly on its axis (once every 243 Earth days). Venus is of a similar size to the Earth, and is expected to have a similar-sized core. A result of Venus' slow rotation, however, is the apparent absence of fluid motions in the core, and production of only a very weak magnetic field (1/25,000th that of the Earth).

Venus has a well-developed ionosphere resulting from the effects of solar radiation on its atmosphere. The ionosphere is involved with the principal interaction between Venus and the solar wind, creating a bow-shock about 2000 km (0.3 Venus-radii) upwind of the planet. Magnetic field lines in the solar wind flowing past Venus become draped around the planet and drawn into a long downwind "tail" structure in a manner similar to those around comets.

At times of intense activity, the solar wind can be driven into the ionosphere of Venus, magnetizing it. Given the absence of a well-developed magnetosphere in which particles can be accelerated, in the near-Venus neighborhood, auroral activity can be dismissed as the cause of the Ashen Light occasionally reported to illuminate the planet's night-side by some visual observers.

Mars

Second-smallest of the terrestrial (rocky) planets, Mars has a diameter of 6800 km, roughly half that of the Earth. As a result of its comparatively small size, Mars has cooled significantly since its formation, and although massive volcanoes like the 24 km high Olympus Mons have been built in the past, Mars' geological activity now appears to be minimal.

As at Venus, the ionosphere of Mars creates a bow shock in the solar wind upwind of the planet. Extensive spacecraft exploration, by orbiter and by landers and rovers on the surface, has shown no sign of an extended global magnetic field. The orbiting NASA Mars Global Surveyor has, however, found regions of anomalous magnetic field in the planet's crust—perhaps remnants of a stronger primordial field.

Observations from the ESA Mars Express spacecraft in August 2004 recorded what appears to have been highly-localized auroral activity, associated with the strongest of the crustal magnetic anomalies. This is found at 117°E, 52°S amidst the heavily cratered ancient terrain of the martian southern hemisphere. Ultraviolet detectors aboard Mars Express recorded a 30-km-wide, 8-km-high patch of auroral emission at an altitude of 130 km in the martian atmosphere above the anomaly. The emission was characteristic of excitation by electrons accelerated in a strong local magnetic field.

It is uncertain whether this glow in the martian night sky would have been visible to a hypothetical observer on the surface. In the absence of a strong global magnetic field, Mars' skies cannot be host to extensive, bright aurorae like those seen by Earth-based observers.

CHAPTER SEVEN

Scientific Investigations

Early observers and theorists of the aurora classed it along with other atmospheric phenomena as a "meteor." In common with many of his other ideas that remained unchallenged until well into the sixteenth century, Aristotle's fourth century BC view of these events being the result of ignition of rising vapors below the innermost celestial sphere prevailed for some time. An alternative, proposed by the Roman philosopher Seneca in his *Questiones Naturales*, was that aurorae were flames viewed through chasmata—cracks in the heavenly firmament.

Even as late as the seventeenth century, no-one was really sure about the aurora's nature. The French mathematician and philosopher Rene Descartes (1596–1650), for example, considered aurorae to originate from sunlight reflected by particles in cirrus clouds.

The Eighteenth Century

By the beginning of the eighteenth century, scientists were actively investigating the terrestrial magnetic field, thanks largely to its importance for navigation at sea. Early in his career, the great English astronomer Edmond Halley (1656–1742) was involved in measuring magnetic dip in the north Atlantic. Following his observation of a major display from London on 16 March 1716, Halley came close to the suggestion that auroral rays were the result of particles flowing along magnetic field lines. Having lived through the Maunder Minimum period of apparently diminished sunspot and auroral activity (Chapter 2), Halley had a long wait for this, his first display:

"Out of what seemed a dusky Cloud, in the N.E. parts of the Heaven and scarce ten Degrees high, the Edges whereof were ringed with a reddish Yellow like as if the Moon had been hid behind it, there rose very long luminous Rays or Streaks perpendicular to the Horizon, some of which seem'd nearly to ascend to the Zenith. Presently after, that reddish Cloud was propagated along the Northern Horizon, into the N.W. and still further westerly: and immediately sent forth its Rays after the same manner from all Parts, now here, now there, they observing no Rule or Order in their rising. Many of these Rays seeming to concur near the Zenith, formed there a Corona..."

[From Halley's "Account of the late surprising appearance of the Lights seen in the air on March 16 last", *Philosophical Transactions of the Royal Society* **29** (1716)].

A further similarly extensive display occurred on 10 November 1719, during which Halley measured the position of the central point of the corona.

A major student of the aurora at this time was the Frenchman Jean-Jacques d'Ortour de Mairan (1678–1771). de Mairan is credited with the first measurements of auroral height, made in 1726. He also noted the seasonal effect in the frequency with which aurorae are detected at lower latitudes, which peaks around the equinoxes. His theories included the suggestion that the solar atmosphere, extending all the way to the Earth, was involved in generating the aurora. A catalog of auroral records was produced by de Mairan, published in his book *Traite Physique et Historique de l'Aurore Boreale* under the auspices of the French Academy of Sciences in 1733.

Around this time, George Graham (1674–1751) in London began taking detailed measurements of the daily fluctuations in the local magnetic field. This work was extended in 1741 by the Swedish physicist, astronomer, and mathematician Anders Celsius (1701–1744) and his assistant Olof Hiorter, who discovered a correlation between days of disturbed magnetic field and auroral activity. Celsius and Graham found that magnetically disturbed days in Uppsala, Sweden, coincided with days which were also disturbed in London.

The first European to observe the Aurora Australis was Captain James Cook, on 17 February 1773, while in the Indian Ocean near latitude 58°S. The display was recorded in HMS *Endeavour*'s log:

"...between midnight and three o'clock in the morning, lights were seen in the heavens, similar to those in the in the northern hemisphere, known by the name of Aurora Borealis..."

These observations confirmed de Mairan's suggestion that there should be aurora around the southern, as around the northern, polar regions.

Among further theories put forward to account for the aurora in the late 18th century was that of Benjamin Franklin (1656–1742), who proposed that it was a lightning effect produced as mobile hot air from the tropics descended into the high polar atmosphere.

The Nineteenth Century

The nineteenth century was a time of rapid improvement in understanding of the aurora, paralleled by advances in other, related areas. Around 1844, the existence of the roughly 11-year sunspot cycle was discovered by Heinrich Schwabe (1789–1875), a German pharmacist and amateur astronomer.

Strong hints to the association between solar activity and enhanced auroral activity must surely have been given following Richard Carrington's 1859 observation of a white-light solar flare and the extensive auroral activity on subsequent nights (Chapter 5).

During the early nineteenth century, it became recognized that aurorae occurred most frequently within certain latitudinal zones, as surmised by the German geographer Muncke in 1833 and the Yale professor Elias Loomis in 1860. Many catalogs of auroral records were compiled by nineteenth century researchers, including one by Hermann Fritz in Zurich. Fritz introduced the term *isochasms* to describe geographical places sharing the same frequency of auroral visibility.

Spectrographic techniques were applied to auroral observation by the Swedish physicist Anders Jonas Angstrom (1814–1874) in 1867, who found the aurora to produce a strong emission in the yellow–green region of the spectrum (later identified as the 557.7 nm "auroral green line" of excited atomic oxygen).

The Twentieth Century

Attempts had been made in the eighteenth and nineteenth centuries to measure the heights of aurorae using the technique of visual triangulation, which was successfully applied to the study of meteors by the German observers Brandes and Benzenberg in 1798. The most prolific series of auroral triangulation observations was obtained by the Norwegian Carl Stormer and his colleagues in a photographic program initiated in 1911.

Photographic emulsions had improved sufficiently by Stormer's time that the images of auroral features obtained by his Norwegian observing network could readily be used for measurements and triangulation. Stormer and his colleagues worked from about 20 observing stations, each separated by at least 20 km, and in telephone contact with each other so that exposures could be made simultaneously. Observing in often extremely cold conditions, they used robust Krogness cameras, which contained a minimum of moving parts which could become frozen or jammed.

Basically, the Krogness camera consisted of a dark chamber to hold a 10×14 cm glass plate, coated with photographic emulsion, and a fast (usually $f/1.5$) cine lens which could be slid to six positions: each plate therefore contained six different exposures. Exposures were made by flipping up a light-tight flap over the lens. The camera swiveled on an altazimuth mount and had a simple cross-wire sight.

Observations by Stormer's network did much to reveal the distribution of activity around the auroral oval. Some 40,000 photographs were taken during this work, enabling over 12,000 accurate auroral measurements to be made.

During the early twentieth century, important developments in methods for observing the Sun were also made. George Ellery Hale and his colleagues at Mt. Wilson Observatory in America devised the spectrohelioscope allowing observation of the Sun in single, isolated wavelengths, in the late nineteenth century. Routine observations of the Sun in the wavelengths of hydrogen-alpha or ionized calcium became possible. Activity in the solar chromosphere was opened to observation, bringing the discovery that this region was far from quiescent! The correlation between flares visible in the wavelength of hydrogen-alpha in the sunspot regions and major aurorae became

Figure 7.1.
Dr David Gavine,
Director of the BAA
Aurora Section,
demonstrates the
operation of a
Krogness camera.

accepted. Further extension of spectroheliographic equipment allowed detection of the magnetism associated with sunspots in 1908.

During the 1930s, the French astronomer Bernard Lyot developed his coronagraph, an optical means of producing an "artificial eclipse" in the telescope, allowing high-altitude observatories to monitor activity in the inner solar corona on a routine basis.

The development of radio communication, starting from Marconi's historic transatlantic transmissions in 1901, led to the discovery of the layers of Earth's ionosphere by the mid 1920s. In 1940, Harang and Stoffregen made the first investigations of radio wave reflection from auroral ionization in the high atmosphere. This period also saw the first detection of radio waves from the Sun (and, indeed, the birth of radio astronomy in general), allowing solar activity to be followed over a still wider range of electromagnetic wavelengths.

Development of Auroral Theory to the 1950s

By the beginning of the twentieth century, the scene was set for major developments in theories to explain the cause of the aurora. The development of models of atomic

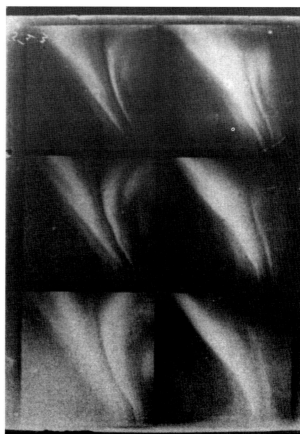

Figure 7.2. A typical Krogness camera plate, with six recorded images of the aurora. Image: BAA Aurora Section archive.

structure, and understanding of the nature of subatomic particles—electrons and ions—brought about important advances. Studies showed that the movement of electrons could be influenced by both electrical and magnetic fields. Extending these findings, the Norwegian physicist Kristian Birkeland (1867–1917) arrived at the suggestion, in 1896, that aurorae are most frequently seen around the polar regions, because it is toward here that electrons from an external source are directed by the Earth's magnetic field, and that the Sun might be a source of streams of such fast-moving electrons.

Like his contemporary Stormer, Birkeland attempted to obtain auroral height measurements. He also conducted laboratory experiments to expand his theories. A magnetized iron ball (called a *terella* by Birkeland), painted with fluorescent material, was suspended in a vacuum chamber and bombarded with electrons ("cathode rays"), and the distribution of the induced fluorescence determined. The results led Birkeland to develop his ideas of how electron beams could stimulate high-latitude aurorae.

The recognition that solar activity in some way influenced terrestrial aurorae was reinforced by the findings of Julius Bartels in the 1930s. Bartels recognized the connection between the 27-day recurrence of geomagnetically active periods and the rotation of the Sun.

Important developments in the theories to account for geomagnetic storms were made by Sydney Chapman and Vincenzo Ferraro in the 1930s. In particular, their work led to the development of ideas regarding the terrestrial magnetosphere.

From further observation, it became apparent that the solar atmosphere as a whole must be continually expanding into interplanetary space, including the immediate environs of the Earth. Ludwig Biermann introduced the idea that the ion tails of comets might provide indirect means of observing variations in the outflow from the Sun. A fuller model for this *solar wind* was proposed by Eugene Parker in 1957, and subsequently confirmed by early spacecraft missions.

Theoretical developments thereafter centered on clarifying the interactions between the solar wind and the magnetosphere that result in the production of auroral activity.

High-Latitude Observations

Collaborative expeditions to high latitudes during the International Polar Years of 1882–1883 and 1932–1933 resulted in collection of a wide range of information on the aurora, along with meteorological and other phenomena in the Arctic. Since the 1930s, establishment of permanent high-latitude auroral observatories has allowed valuable work to be carried out from locations such as College, Alaska, Tromso and Oslo in Norway, and Kiruna in Sweden. The relatively dependable occurrence of auroral activity at these locations allowed more intensive study with advanced equipment. Fast "all-sky" cameras could be used to monitor the distribution and form of aurora through the course of the nights. Detailed spectroscopy was carried out using diffraction gratings, whilst radar observations probed the ionosphere.

Sounding rockets allowed direct investigations of atmospheric conditions at high altitudes during aurorae. Black Brant rockets, capable of reaching altitudes up to about 900 km, have been launched for auroral study from the Fort Churchill Range in Canada, the Poker Flats Range in Alaska, and elsewhere. Launches have also been made from Kiruna, and other European locations.

Much of this more advanced scientific equipment was in place in time for the International Geophysical Year of 1957–1958, which was to yield a great deal of useful information relating to auroral phenomena and their causes.

The International Geophysical Year and Beyond

Timed to coincide with the maximum of sunspot cycle 19—which turned out to be the most active on record, with many memorable mid-latitude aurorae—the International Geophysical Year (IGY), which ran for the 18 months between July 1 1957 and December 31 1958, saw international collaboration in auroral and atmospheric study on an unprecedented scale. The scientific program demanded continuous visual, photographic, and radio monitoring of solar activity and study of its subsequent effects—including auroral activity—on the terrestrial environment. Meteorological and oceanographic observations also had a role in the overall collection of data for the IGY: basically, the aim was to gather as much information on a wide range of interrelated phenomena from as many geographical locations as possible. An important aspect of the IGY was the extension of the observational network to the southern polar region; the Polar Years had involved research solely in the Arctic.

The value of visual auroral observations was recognized, and amateur astronomers were actively encouraged to participate by taking measurements and providing descriptive accounts of activity. Reports of this nature, made following standard guidelines, were also collected from meteorological observers, sailors and aircrew, considerably extending the geographical range covered.

Professional observers operated a network of all-sky cameras around the polar regions in order to take simultaneous images of the aurora's distribution. Spectrographic analyses were also carried out, and radar techniques applied to the study of ionospheric movements during auroral activity.

Balloons and rockets were used for high altitude cosmic ray measurements. Sounding rockets—among them the British Skylark, capable of reaching heights up to about 210 km—were launched into regions of the atmosphere affected by auroral displays to obtain *in situ* measurements of particle energies and densities. America had ambitious plans to launch several artificial satellites into Earth orbit to supplement other IGY activities. The first successful American launch, the 14 kg Explorer 1 satellite which entered Earth orbit on 31 January 1958, was significant in carrying instruments capable of measuring its local radiation environment. Using this equipment, the Van Allen belts were discovered and the detailed exploration of the near-Earth space environment had begun. Over the following six months, two more Explorer satellites were launched.

Like so many other scientific ventures, IGY did not yield its full fruits until some years after the observations—an immense collection of datahad been made. An important finding, resulting from analysis of all-sky photographic observations, was the conclusion by Feldstein and Koroshove in 1963 that the aurora is distributed in oval regions around either geomagnetic pole.

The success of the IGY was followed by the International Years of the Quiet Sun (IQSY) in 1964–1965, intended to study the same phenomena under the conditions prevailing at sunspot minimum.

Auroral studies continue from an extensive network of ground-based observatories. The equipment available grows ever more sophisticated, but observations are still supplemented by photographic and, to a lesser degree, visual means. Low-light television cameras allow recording of the often rapid motions of auroral forms at high latitudes; some of the spectacular results have been made available as commercial videos for home consumption.

Many observing stations for auroral and other research now operate in the Antarctic. The Japanese-funded Syowa station is well placed for auroral observations. Halley Base, part of the British Antarctic Survey, is ideally located for studies of auroral effects at the boundary of the plasmasphere. Radio investigations, including work with the Advanced Ionospheric Sounder (AIS) and assessment of ionospheric particle densities using a riometer relative ionospheric opacity meter are carried out from Halley.

Sounding rockets still provide the means of briefly sampling conditions in the auroral layer. Facilities such as EISCAT also allow more detailed study of ionospheric processes during aurorae. EISCAT (European Incoherent Scatter facility) is a collaborative venture between several European countries. UHF radio signals broadcast from a 32-meter dish at Tromso in Norway, and reflected from auroral structures, can be detected by dishes at Kiruna (Sweden) and Sodankyla (Finland), and at Tromso itself. VHF signals are also transmitted from, and received back at a 120 by 40 meter "trough" at Tromso. Measurements allow ionospheric particle densities and movements to be assessed. EISCAT was extended by the addition of facilities at Spitsbergen in 1996. A further addition to the network of HF radars is the Co-operative UK Twin Located Auroral Sounding System (CUTLASS) at stations in Finland and Iceland, which commenced operation in 1995. Funded by the UK, Finland and Sweden, CUTLASS is designed to monitor ionospheric particle movements.

Studying the Solar Wind and Earth's Magnetosphere

As well as investigating the magnetic environments of the major planets, spacecraft have, since the early 1960s, opened up a vastly greater understanding of the interplanetary medium. Magnetometer equipment aboard the Mariner 2 mission to Venus in 1962 confirmed the existence of the solar wind predicted by Eugene Parker. A series of Interplanetary Monitoring Platform (IMP) probes yielded further valuable information on the solar wind and, of course, the Pioneer and Voyager spacecraft have extended our knowledge of the heliosphere almost out to its far-distant boundary.

Prior to the ESA Ulysses mission, launched from the Space Shuttle *Discovery* in October 1990, measurements of the solar wind had been, of necessity, confined to more or less the ecliptic plane. Following encounter to within 450,000 km of Jupiter in February 1992, Ulysses was flung into a high-inclination orbit around the Sun, passing over the southern pole from late June to early July 1994, and reaching a maximum ecliptic latitude of 80.2°S on 13 September 1994. Between June and September of the following year, Ulysses was at high northerly ecliptic latitudes.

Ulysses carries instruments for measurement of solar wind ion composition, and with which to examine interactions between cosmic rays and the solar wind. Magnetometers measure the interplanetary magnetic field, and readings are also taken of radio and plasma wave phenomena. Much as expected, Ulysses has found that the solar wind emerges at a higher velocity at high ecliptic latitudes.

A further passage to high southerly ecliptic latitudes occurred in November 2000 (this time close to the maximum of the sunspot cycle) with a corresponding northerly passage from September to December 2001. Ulysses' third passage over the south of the Sun in November 2006, by contrast, coincides with sunspot minimum at the end of cycle 23.

Closer to home, spacecraft investigations have also improved our understanding of magnetospheric conditions and processes involved with the aurora. A ground-based

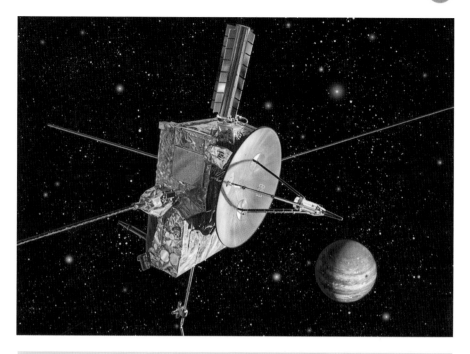

Figure 7.3. Artist's impression, by David Hardy, of the ESA Ulysses probe. Image: ESA.

observer (or even a satellite in low Earth orbit) can see only a small part of the auroral oval. Satellites in highly elliptical polar orbits, from which an entire hemisphere of the Earth can be viewed at apogee, can, however, allow study of global-scale phenomena in the auroral oval. The Canadian ISIS-II satellite, launched in 1970, was the first to provide whole-oval images.

Perhaps most widely known are the images taken from the NASA Dynamics Explorer-1, launched in August 1981 as one of a pair of satellites. The eccentric orbit of Dynamics Explorer-1, with apogee at 22,000 km (3.5 Earth-radii), allowed the whole auroral oval to be visualized for intervals of up to 5 hours—long enough to follow the progress of substorms, for example.

The sister-satellite Dynamics Explorer-2 was placed in a lower, more circular orbit and provided complimentary particle and other measurements before re-entering Earth's atmosphere in February 1983.

Ion releases from satellites have been used to map Earth's magnetosphere. A multi-satellite mission, AMPTE (Active Magnetospheric Particle Tracer Explorer), jointly funded by the US, UK, and Germany, was launched in 1984. This carried out a series of barium and lithium releases, detection of which at various magnetospheric locations by component satellites helped improve models of particle movements. Just after Christmas 1984, a release from AMPTE produced an "artificial comet" in the solar wind. These experiments were followed in 1991 by the CRRES (Combined Release and Radiation Effects) satellite, which carried out further gas releases.

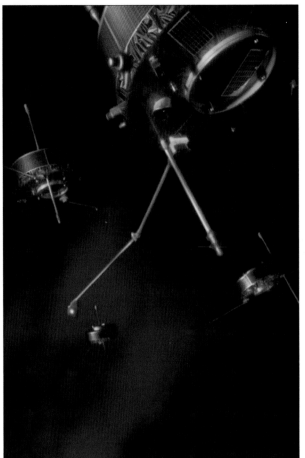

Figure 7.4. Artist's impression of the four Cluster spacecraft in formation Image: ESA.

Scientists long ago realized that a multiple-spacecraft mission offers the most effective means of studying large-scale field changes and particle motions in the magnetosphere. This goal is being achieved through the ESA Cluster mission, a set of four identical 2.9 meter diameter by 1.3 meter high satellites (the individual components are named Rumba, Tango, Salsa, and Samba), which are flown in a tetrahedral formation enabling simultaneous measurements in three dimensions from different magnetospheric locations. Following the catastrophic loss of the Ariane 5 rocket on the initial launch attempt in June 1996, the rebuilt Cluster mission was successfully placed in orbit using two modified Soyuz launchers in July and August 2000.

The Cluster satellites are in a 57-hour elliptical polar orbit with perigee 17,200 km (2.7 Earth-radii) above Earth, and apogee at 120,600 km. Distances between the individual components vary between 125 and 2000 km, and each is equipped with a suite of 11 detectors. The satellites each have four 50-meter and two 5-meter instrument-tipped booms. The Cluster mission has provided valuable insights to how the magnetosphere responds as coronal mass ejections arrive, and has confirmed the occurrence

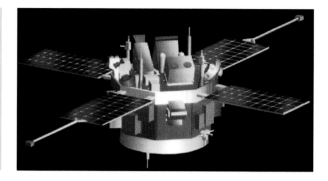

Figure 7.5. The Advanced Composition Explorer spacecraft. Image: NASA.

of plasma waves at the magnetopause. A surprising discovery has been the rapid movement of the polar cusps in response to changes in the solar wind.

Previously scheduled to operate from February 2001 to December 2005, the Cluster mission has been extended to the end of 2009.

Occupying a "halo" orbit around the L1 Lagrangian point 1.5 million km "upwind" from Earth in the direction of the Sun, NASA's ACE (Advanced Composition Explorer) spacecraft takes measurements of conditions in the solar wind, about an hour before it impacts on the terrestrial magnetosphere. Launched in August 1997, ACE carries instruments to measure energetic ions and electrons, magnetic fields, high-energy-particle fluxes and composition of the solar wind. Data returned by ACE provide valuable early warnings of potential geomagnetic storm conditions. Although comparatively small (1.6 meter diameter, 1.0 meter high), the solar panel-powered craft has sufficient propellant onboard to maintain operation around the L1 point until about 2019.

Measurements of the solar wind just upstream of the magnetosphere were taken prior to ACE by the WIND spacecraft, launched in 1994—one of several missions in the International Solar Terrestrial Physics Programme (ISTP), which also involved ground-based observations. Among other ISTP missions, FAST (Fast Auroral Snapshot Explorer) obtained field and plasma measurements in the auroral regions. Polar, launched into a high-inclination orbit in 1996, has picked up where Dynamics Explorer-1 left off, obtaining images of the auroral ovals in visible, UV, and X-ray wavelengths, along with particle measurement, in a mission extended to 2007.

Launched in March 2000, NASA's IMAGE (Imager for Magnetosphere-to-Aurora Global Exploration) measured plasma populations and interactions in the magnetosphere from a 14.2-hour polar orbit with apogee at 45,920 km (7.2 Earth-radii), perigee 1000 km. Among its many discoveries was the distortion of the magnetosphere following the "Hallowe'en Storms" in 2003 (Chapter 5). Contact was lost with the satellite—which far surpassed its nominal two-year lifetime—in December 2005.

Staring at the Sun

Solar astronomy—a crucial link in unraveling the auroral mechanism—has also benefited greatly from advances in satellite technology. Ground-based observatories such as those at Kitt Peak, Big Bear, and La Palma, are eminently capable of routine coverage of the Sun's activity in many electromagnetic wavelengths. Observations at X-ray and ultraviolet wavelengths, however, were once restricted to fleeting glimpses from rockets making brief flights above most of the obscuring atmosphere.

Satellites provided the opportunity to make longer-duration observations, unimpeded by atmospheric absorption, at the wavelengths strongly emitted by energetic solar phenomena. A series of Orbiting Solar Observatories (OSO 1–8) was launched by NASA, followed by the highly successful manned Skylab space-station in 1973–1974.

Skylab included a number of instruments for solar observation, notably the Apollo Telescope Mount. The Skylab missions, three in all lasting a total of 171 days, coincided with the declining phase of solar cycle 20, which reached its minimum in 1976. X-ray observations from Skylab confirmed the existence of late-cycle coronal holes. Two persistent coronal holes were observed in detail from Skylab, including the "Boot of Italy" feature, so named for its profile against the solar disk. Skylab also revealed the existence of X-ray "hot-spots" (apparently coincident with the bases of long streamers in the solar corona) and of short-lived X-ray bright features.

These observations were followed up by another successful satellite, the Solar Maximum Mission (SMM, or "Solar Max"), launched in February 1980. As its name suggests, SMM was aimed primarily to study the Sun at a different phase of the sunspot cycle from that observed during the Skylab missions. After a few months of virtually perfect operation, however, many of the detectors aboard SMM were disabled by an electrical fault in April 1981. Repairs were carried out in orbit by astronauts aboard the Space Shuttle *Challenger* in 1984, in time for SMM to cover the *minimum* between cycles 21 and 22 during 1986 in great detail, before the increased atmospheric drag resulting from intense solar flare activity early in cycle 22 brought about its demise.

SMM obtained detailed observations of solar flare and other phenomena in X-ray and ultraviolet wavelengths. Coronagraph observations from SMM resulted in the discovery of several comets, as did the Solwind instrument aboard a US Military satellite, P78–1.

The Skylab and SMM studies of the Sun were followed in the early 1990s by the immensely successful Yohkoh (Japanese for "Sunbeam") orbital solar observatory. A joint Japanese, British, and American venture, Yohkoh was launched on a planned 3-year mission in August 1991, carrying instruments the detection of low- and high-energy (soft and hard) X-rays, and gamma rays. In the end, Yohkoh operated flawlessly until electrical failure in mid-December 2001, amassing just over 10 years' worth of data graphically demonstrating the solar corona's changing X-ray output over the sunspot cycle. Yohkoh was destroyed on re-entering Earth's atmosphere in 2005.

A mainstay of solar monitoring is the NASA/ESA SOHO (Solar and Heliospheric Observatory), launched in December 1995. Like ACE, SOHO is in a "halo" orbit around the L1 point upwind of Earth towards the Sun. SOHO instruments have been important to helioseismological studies, but for those with an interest in solar–terrestrial relations it is perhaps the coronagraphs that merit most attention. The LASCO (Large Angle and Spectrometric Coronagraph) instruments monitoring the inner soar atmosphere have provided stunning views of CMEs, demonstrating that the corona is, at times, a very dynamic environment. The C3 coronagraph, with a field of view imaging to 32 solar radii, has been the source of graphic time-lapse movies which are available in the public domain at the SOHO websites (Chapter 3).

Figure 7.6. Artist's impression of the SOHO spacecraft, which provides continuous monitoring of solar activity. Image: NASA/ESA.

Among its other instrumentation, SOHO has an extreme ultraviolet telescope (EIT) and magnetographic equipment (MDI).

SOHO has had a checkered history. After two and a half years of operation, contact was lost on 24 June 1998 as a result of a software problem. The spacecraft ended up with its solar panels pointed away from the Sun, and its hydrazine fuel—used for maneuvering—frozen. Engineers were able, gradually, to re-orient SOHO, and by 2 February 1999 it was again fully operational. It has worked well ever since—so well, in fact, that future missions such as STEREO are actually being designed around SOHO.

Launched in April 1998, TRACE (Transition Region and Coronal Explorer) is another NASA satellite, which is designed to obtain near-simultaneous observations of phenomena in the inner solar atmosphere at a range of spectroscopic wavelengths. Results from TRACE have helped clarify coronal heating processes. Observations were coordinated with those from SOHO during the rising phase of sunspot cycle 23.

NASA's STEREO (Solar TErrestrial RElations Observatory) mission, scheduled for launch in the second half of 2006, will use two nearly identical spacecraft to provide, for the first time, a three-dimensional view of coronal mass ejections. The spacecraft will orbit the Sun at essentially the same distance as Earth, but one will be ahead of the planet, the other behind. This arrangement will afford views some way around the solar limb in comparison with the terrestrial perspective. The two 620 kg spacecraft are equipped with coronagraphs and particle detectors, and will be able to triangulate the positions of interplanetary radio bursts. Observations will be coordinated with those from SOHO, and the STEREO mission is expected to last two years.

The solar-terrestrial link is becoming ever better understood through the use of scientific tools of which our forebears can scarcely have dreamed. In the five decades since the International Geophysical Year, scientists have developed a much more complete picture of auroral phenomena and their underlying causes. Much remains to be clarified in detail and—as discussed in Chapter 1—our increasing reliance on satellite technology has made Space Weather a priority topic.

Marking the 50th anniversary of the IGY, 2007 has been adopted as the International Heliophysical Year, during which large-scale research collaborations will hopefully help unravel some of the mysteries of Space Weather (http://www.ihy2007.org). In contrast with the situation in 1957–1958, amateur astronomers are unlikely to be able to contribute a great deal observationally: satellites can now far more accurately pinpoint the location and movements of the auroral ovals. This does not mean that amateur observers should abandon all interest in the aurora! Ron Livesey and others quite correctly point out that, while they are unlikely to lead to new discoveries, modern amateur visual and photographic observations of the aurora provide continuity with those in the past. It's perhaps still valid to compare visual observers' impressions of major events like the 1989 or 2003 storms with those reported in earlier times, whatever the more advanced satellite and other measurements may show. Equally to the point, as with observation of, say, deep sky objects or total solar eclipses, the fact that little, if any, original science is likely to emerge should not be allowed to detract from the simple pleasure and satisfaction to be gained from witnessing an astronomical spectacle for oneself.

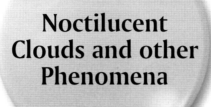

CHAPTER EIGHT

Noctilucent Clouds and other Phenomena

Noctilucent Clouds

At those latitudes most favored for observations of aurorae, the summer night-time sky never becomes completely dark. In northern Scotland (59°N) at midsummer (21 June), for example, the Sun never sinks more than about 9° below the northern horizon. Under such circumstances, twilight persists all night, and only the brighter naked eye stars are readily visible: astronomical observations are difficult, and this might well be regarded as "off-season" for aurorae, which will normally be swamped by the bright sky. There is, however, one high atmosphere phenomenon of which observations can usefully be made during the twilit summer nights at higher latitudes: *noctilucent clouds.*

While they are of a very different nature to the aurora, the occurrence of noctilucent clouds (NLC) is apparently influenced by the level of solar activity, and occurring at a level in the high atmosphere not far below that where aurorae occur, they have become a subject of attention for aurora observers. There are some similarities in the way in which observations are recorded.

It may seem unusual for those normally engaged in astronomical work to actively look for cloud phenomena, but noctilucent clouds are quite different from the lower, tropospheric, clouds which cause so much frustration to observers on the ground. While the highest tropospheric clouds (the "mares' tails" of cirrus, among which the optical phenomena discussed in Chapter 1 may be seen) reach a maximum altitude of about 15 km in the atmosphere, noctilucent clouds are formed in thin sheets at altitudes in excess of 80 km. Since there is very little water vapor present in the atmosphere at such great heights, noctilucent clouds are extremely tenuous, and may be seen only under certain illumination conditions; they are lost in the bright daytime sky, only

Figure 8.1. Type III noctilucent cloud display imaged from Sussex by the Author on 29–30 June 1994.

becoming visible as twilight deepens after sunset. In order for noctilucent clouds to become visible, the Sun must lie between 6° and 16° below the observer's horizon: too high, and the sky will be over-bright; too low, and the noctilucent clouds themselves will be in the Earth's shadow. Figure 8.2 illustrates the illumination conditions necessary for noctilucent clouds to be observable. The sky must be sufficiently dark for the noctilucent clouds to appear bright by contrast, while the clouds themselves remain in sunlight above the Earth's shadow. Looking up through the Earth's shadow, the observer should see any foreground tropospheric clouds in darkness against the noctilucent cloud field over the poleward horizon.

Noctilucent clouds are believed to form only during the summer months in either hemisphere. The noctilucent cloud "observing season" in northwest Europe and Canada extends from late May to early August (perhaps peaking in late June and the first week of July), while southern hemisphere observers are favored in December and January. Records of noctilucent cloud sightings in the southern hemisphere are rather

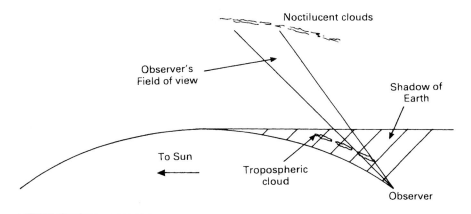

Figure 8.2. Noctilucent clouds become visible at high temperate latitudes during the summer months by virtue of their great height (in exces of 80 km). Noctilucent clouds are sufficiently high to remain in sunlight after tropospheric clouds are in darkness in Earth's shadow. In this figure, the curvature of the Earth is exaggerated.

sparse by comparison with those from the northern hemisphere, thanks mainly to a less favorable disposition of the main land-masses there relative to the main zones of occurrence. Reports have been made from Davis Station ($68°35'$S, $77°58'$E) in Antarctica.

Noctilucent clouds usually form between latitudes of about $60°$ and $80°$, and may be observed to as low as about $50°$ latitude. In contrast with the aurora, the regions of most frequent noctilucent cloud occurrence are determined principally by geographical, rather than geomagnetic, latitude. Observations are most frequently made from those latitudes where the summer night sky is reasonably dark, and the cloud-sheet may be observed at an acute angle; Aberdeen ($58°$N) and Helsinki ($61°$N) are favored in this respect, for example.

During the spring and summer months at high latitudes, upwelling of cold, moist polar air is thought to carry water vapor to the extremely high altitude of the mesopause. Average temperatures at the mesopause are low, on the order of $-108°$C. Temperatures as low as $-162°$C have been measured during rocket flights into noctilucent cloud fields.

At the mesopause, water vapor may condense around small nuclei to form noctilucent clouds. The precise nature of the condensation nuclei remains subject to debate, despite attempts to recover noctilucent cloud particles using "Venus Flytrap" sounding rockets over Sweden in the 1960s, and further rocket flights into noctilucent clouds over Sweden and Canada up to the early 1970s. Direct spectroscopic examination of noctilucent clouds from ground level is rendered impossible by the bulk of atmosphere between observer and target.

A logical source of condensation nuclei might be meteoric debris: meteors become luminous at heights of about 100 km above the Earth's surface, and residual particles from each meteor's ablation should remain suspended in the high atmosphere for periods estimated at about three years. Volcanic material injected into the stratosphere during violent events such as the Krakatoa eruption in 1883 or Mount Pinatubo in 1991 may also be carried aloft to provide a source of condensation nuclei (and, possibly, water vapor) for noctilucent clouds. There are strong indications, however, that volcanic activity may also suppress noctilucent cloud formation; a relative dearth of sightings in 1992 has been attributed to disruption of the upwards transport of water vapor in the atmosphere by material ejected by the previous year's Mount Pinatubo eruption in the Philippines.

Historically, noctilucent clouds were first observed in the years immediately following the Krakatoa explosion, perhaps lending credibility to the possibility of volcanic sources of nuclei. It has, however, also been suggested that the increased prevalence of spectacular sunset and twilight optical phenomena following the global dispersal of the dust ejected by the Krakatoa explosion may have led to improved observational vigilance at those times of night when noctilucent clouds might be expected to become visible, thereby accounting for their initial detection in the mid-1880s.

Photoionization of atmospheric particles by solar ultraviolet radiation may provide another possible source of condensation nuclei for noctilucent cloud particles.

The source of the water vapor widely believed to comprise the bulk of noctilucent clouds is also subject to debate. One theory put forward by Gary Thomas (University of Colorado) and co-workers in 1989 suggests that methane released into the atmosphere by industrial activities such as oil exploration and coal mining may rise into the stratosphere, there to be dissociated by solar ultraviolet radiation above the ozone layer with subsequent chemical reactions resulting in formation of two molecules of

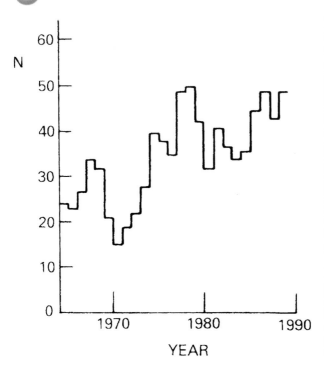

Figure 8.3. The frequency with which noctilucent cloud displays have been visible from northwest Europe has shown an apparent gradual increase since the 1960s, on which is superimposed the solar cycle effect. Reproduced from Gadsden, M., J. Atmos. Terres. Physics **52** 247–251 (1990) by permission of the author.

water per methane molecule as a by-product. Carried higher into the atmosphere, the water could condense to form noctilucent clouds. This theory may also account for an apparent dearth of noctilucent cloud sightings in records prior to the late nineteenth century, there previously having been insufficient quantities of water vapor present in the high atmosphere to allow cloud formation. Increasing liberation of methane with time has been postulated as leading to an increased frequency of noctilucent cloud displays and a parallel increase in the average brightness of those displays. Some observations from northwest Europe suggest that the northern hemisphere noctilucent cloud observing season may have lengthened during the late 1980s, with sightings being made earlier into May and later into August than previously. Further observations are required in order to assess the reality of these suspected changes.

Figure 8.3 summarizes total numbers of noctilucent cloud sightings from northwest Europe over a 24-year period. The gradual long-term increase in noctilucent cloud frequency is apparent, as are superimposed decreases in frequency which appear to be correlated with periods of high sunspot activity in 1970 and 1980.

Condensation of water vapor around the nuclei, whatever their nature, appears to begin at an altitude of about 85 km. As the particles grow in size and mass, they begin to fall under gravity, reaching a maximum density about 82 km altitude. Below this level, the atmospheric temperature begins to rise, and the ice evaporates. Consequently, noctilucent clouds appear in thin sheets.

Noctilucent clouds are in some respects frustratingly located for scientists wishing to study them; several have commented that NLCs are "too high for balloons, too low for satellites." In other ways, NLCs are too high to concern conventional meteorologists, but are patently atmospheric rather than astronomical in nature (though not, of course, likely origin): their study is best embraced by the field of aeronomy, sort of halfway house!

Strictly, NLCs are not really too high for useful study from satellites. Sightings have been made by cosmonauts aboard Mir, and more recently by NASA astronaut Don Pettit aboard the International Space Station. A NASA satellite dedicated to the study of noctilucent clouds (also known by atmospheric scientists as polar mesospheric clouds; PMCs) launched on 29 in September 2006. AIM (Aeronomy of Ice in the Mesosphere) is designed to take measurements over a two-year mission from a 550 km altitude orbit. Instruments will take images of NLCs/PMCs, measure temperatures and water vapor content in the mesosphere (between 50 and 85 km above Earth's surface) and monitor the influx of meteoric dust.

Noctilucent Clouds and Solar/Auroral Activity

Forming in a layer of the atmosphere not far below the region where the aurora occurs, noctilucent clouds might be expected to show some behavioral correlation with solar–terrestrial activity. It was once presumed that auroral activity in the overlying thermosphere should sufficiently warm the mesopause that any noctilucent clouds present at the time would be evaporated. Thus, it is expected that the maximum frequency of noctilucent cloud occurrence should be observed at sunspot minimum, when aurorae are uncommon, while few noctilucent cloud displays should be seen around sunspot maximum.

Observations suggest a more complicated interrelation between the two phenomena. Certainly, large numbers of noctilucent cloud displays were observed from northwest Europe in the solar minimum summer of 1986, including a very bright and extensive display on 23–24 July (Figure 8.4). During the following year, auroral

Figure 8.4. Rarely, noctilucent cloud displays can be extremely bright and extensive. One such instance was the major occurrence of 23–24 July 1986, seen widely across the British Isles, and recorded here from Edinburgh. Image: Dave Gavine.

activity stayed low, and another summer with numerous noctilucent cloud displays followed in 1987. The summers of 1988 and 1989, by which time auroral activity was significant, also presented large numbers of noctilucent cloud displays, however, casting some doubt on the reality of the inverse correlation between noctilucent clouds and aurorae. The frequency of noctilucent cloud sightings from northwest Europe did, however, show a substantial drop in the early 1990s. There are some suggestions of a 2-year phase lag between auroral activity and the inverse-correlated frequency of noctilucent clouds. The summer of 1995, marked by very low sunspot activity, was again graced with a great many noctilucent cloud displays. High solar activity in the early 2000s again saw some reduction in noctilucent cloud sightings, but the northern summers of 2005 and—particularly—2006 were notably productive.

Yet further complications are introduced by simultaneous sightings of noctilucent clouds with a backdrop of auroral activity! Clearly, observations of both phenomena during the summer months will be of increasing value in attempts to unravel the complex interactions taking place in the outer fringes of the Earth's atmosphere.

It is quite likely that the frequency of noctilucent cloud formation is influenced more by the overall level of solar activity at X-ray and ultraviolet wavelengths, than by auroral activity *per se*. The X-ray and UV flux from the Sun plays a significant role in heating of the upper atmosphere. Active regions emitting this short-wave radiation need not be in the correct (geoeffective) alignment for the production of

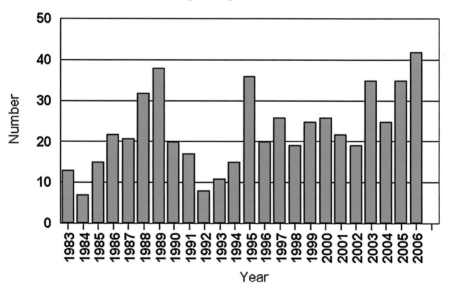

Figure 8.5. Annual frequency of noctilucent cloud displays from the British Isles between 1981 and 2006. The summers of 1995 and 2006, close to sunspot minimum, were notable for large numbers of displays, while sunspot maxima around 1990 and 2000 show a reduction in noctilucent cloud frequency.

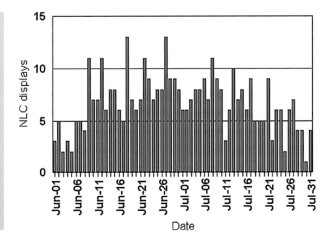

Figure 8.6.
Frequency of noctilucent cloud sightings by date from the British Isles, 1981–2006. The three-week interval centred on the Summer Solstice (21 June) appears to be particularly favoured.

auroral activity in order to have an effect; flares occurring close to the solar limb are as efficient in heating the upper atmosphere as those close to the Sun's central meridian, though only the latter may be followed by auroral activity.

Figure 8.5 presents a summary of total noctilucent cloud sightings from the United Kingdom between 1981 and 2006. The rising frequency toward the end of the period coincides with falling sunspot numbers and the approach of solar minimum. Figure 8.6 shows sightings of noctilucent clouds over these years as a function of date. In general, the best time to see noctilucent clouds from the British Isles and North America appears to be in the last 10 days of June, and into the first week of July.

Appearance and Behavior

Noctilucent clouds bear a superficial resemblance to cirrus, but closer examination reveals diagnostic differences in appearance between the two cloud forms. To begin with, by the time noctilucent clouds are becoming visible on a summer evening, cirrus clouds lying in the same part of the observer's sky should be entering the Earth's shadow and becoming dark. Noctilucent clouds often show a distinctive delicate silvery-blue color, shading off to gold as a result of atmospheric reddening near the horizon. Highly banded structures are common in noctilucent clouds, and considerable fine detail, including a characteristic "herring-bone" pattern, is often revealed by binocular examination. The westwards drift of noctilucent cloud features contrasts with the typical eastwards movement of tropospheric weather systems at the latitudes where noctilucent clouds may be observed.

Noctilucent clouds first become visible once the Sun has set below 6° under the observer's westward horizon, and at this time may cover much of the sky at higher-latitude locations. As the depression of the Sun below the horizon increases toward midnight, the noctilucent cloud field will fade somewhat, and diminish in apparent extent. Where the Sun reaches more than 16° below the horizon, noctilucent clouds will fade out altogether. The brightest area in a display normally lies directly above the location of the setting Sun, and moves along the poleward horizon as the

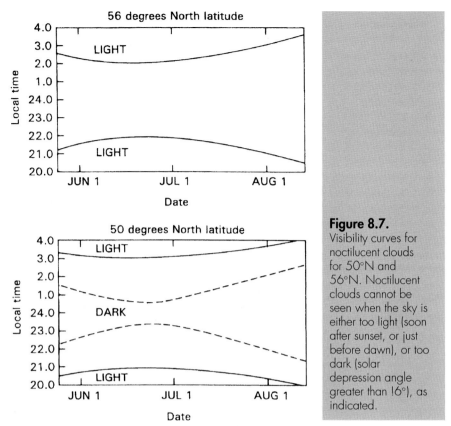

Figure 8.7.
Visibility curves for noctilucent clouds for 50°N and 56°N. Noctilucent clouds cannot be seen when the sky is either too light (soon after sunset, or just before dawn), or too dark (solar depression angle greater than 16°), as indicated.

night goes on. With the approach of dawn, the display can again become more extensive as the Sun rises. When the Sun is less than 6° below the eastern horizon, the sky becomes too bright, and noctilucent cloud forms will appear to melt into the background.

Figure 8.7 provides visibility curves for noctilucent clouds during their summer "season" at latitudes of 50°N and 56°N, corresponding roughly to southern England or the northern United States and Canada, and the more favorably placed central Scotland and southern Scandinavia respectively.

While occasional large displays of noctilucent clouds can fill the whole sky at higher latitudes, the brightest regions of the cloud field normally appear in the sunwards half of the sky. The particles that comprise noctilucent clouds are efficient in scattering sunlight forwards, so that any cloud forms overhead and in the half of the sky opposite the Sun will appear rather faint.

Parallactic photography of noctilucent clouds has been carried out by a number of groups, including Scottish amateur astronomers under the guidance of the late Dr. Michael Gadsden of Aberdeen University. Examination of the simultaneous photographs obtained during such work allows precise measurements to be made of noctilucent cloud movements. Noctilucent clouds are carried westwards by a high atmospheric circulation at velocities of up to 400 km/hr. This circulation may have its origin in collisions between particles carried along in the auroral electrojets (Chapter 2) and particles in the low ionosphere. The same high-atmosphere winds may occasionally be seen distorting the ionization trains left by bright meteors at this altitude.

Visual Observation

Simple observations of noctilucent clouds remain of interest and value to scientists studying processes in the high atmosphere. In particular, amateur astronomers appear to have collected the only significant numbers of systematic noctilucent cloud observations throughout the 1970s and early 1980s. These observations may prove their value in assessing whether long-term changes in the frequency of noctilucent clouds really have occurred over this period. This work, in turn, may have important implications with respect to possible man-made changes in the lower atmosphere.

The Balfour Stewart Laboratory of Edinburgh University was, for many years, the main receiving center for European noctilucent cloud observations. On the Laboratory's closure, responsibility for collecting such observations passed to the Aurora Section of the British Astronomical Association, which in turn provides the observations for archiving at Aberdeen University. Annual reports of all northwest European observations up to 1992 were presented in *Meteorological Magazine,* and are now published in the *Journal of the British Astronomical Association.* Among the active groups contributing data are observers in Denmark, who are ideally placed to record noctilucent clouds. Finnish observers are similarly well positioned to see both aurorae and noctilucent clouds, and summarize their results in the magazine *Ursa Minor* published by the Ursa Astronomical Association. Starting in the late 1980s, systematic records of noctilucent clouds visible from Germany have been collected by the Arbeitskreis Meteore (AKM).

In Canada and the northern United States, observers have become increasingly aware of the importance of recording noctilucent cloud sightings, through the NLC CAN-AM network organized by Mark Zalcik. Annual summaries of North American observations are published in the *Climatological Bulletin.*

Since 1996, an excellent Noctilucent Cloud Observers' Homepage has been run by Tom McEwan, an experienced amateur observer of aurorae and NLC based in Ayrshire, Scotland (http://www.nlcnet.co.uk). Internet-connected observers post sightings and images here, and this has allowed rapid dissemination of information on the visibility of displays. Useful background information, and instructions on NLC recording, can also be found here.

To be of use, observations should be made by standardized methods. As with the aurora, there is a range of typical forms shown by noctilucent clouds, which can be concisely described in observer reports. Figure 8.8 shows the typical noctilucent cloud structures; standard descriptions used by the British Astronomical Association are listed in Table 8.1.

The most fundamental report which can be made is to note that noctilucent cloud was visible from a given location on a particular night. As with auroral observations, the latitude and longitude of the observing station should be given. Double dates should be used to avoid ambiguity—for example, 27–28 June, meaning the night of

Type II   Type IV

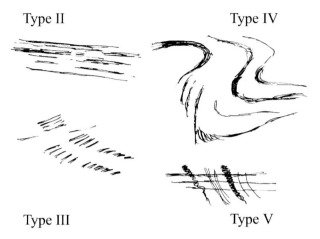

Type III   Type V

Figure 8.8.
Schematic representation of the principal noctilucent cloudforms. Type I is a featureless veil. See also Table 8.1.

June 27 to the morning of June 28. Universal Time (UT = GMT) should always be used for both noctilucent cloud and aurora observations.

Noctilucent cloud displays tend to change less rapidly than aurorae, so that records of the display's condition at 15 minute intervals are normally sufficient. In order to maximize the numbers of directly comparable reports of a given display, records should

Table 8.1. Noctilucent Cloud Forms and Brightness Scale

FORMS

Type I Veil—a structureless sheet, often a background to other forms.

Type II Bands—linear structure, or streaks, parallel or crossing at small angles:
    IIa Bands have diffuse, blurred edges
    IIb Bands have sharply defined edges

Type III Waves—"herring-bone" pattern, resembling sand-ripples on a beach at low tide:
    IIIa Short, straight narrow streaks
    IIIb Long waves and undulations

Type IV Whirls—looped or twisted large-scale structures:
    IVa small angular radius (0.1–0.5°)
    IVb simple curve(s) angular radius 3–5°
    IVc large-scale whirls

Complex: Type O Not described by any of I-IV
    Type S Bright "knots"
    Type P Billows crossing a band
    Type V Interwoven "net" structure

BRIGHTNESS

1: Very weak—barely visible
2: Clearly detected—low brightness
3: Clearly visible—good contrast with twilit sky
4: Very bright—noticeable even by casual observers
5: Extremely bright—may illuminate objects facing the display

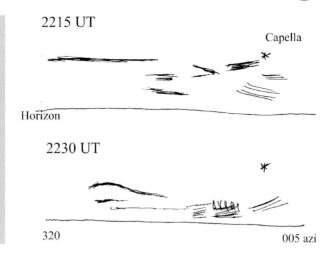

Figure 8.9.
Sketches of the 6–7 July 2005 noctilucent cloud display (see Table 8–2) from the Author's observing log.

ideally be taken on the hour, half hour, and intermediate quarter hours. Observers should note which types of noctilucent cloud structures are present at a given time, and make estimates of the forms' extent in altitude and azimuth. A simple alidade may improve the accuracy with which such measurements can be made. Annotated sketches are often useful in clearly indicating the behavior of a display. Fig. 8.9 gives an example from the author's observing logbook. Table 8.2 is the written report for the same event.

Estimates of the brightness of individual features may be helpful to identification when attempts are subsequently made to combine noctilucent cloud observations from several locations. The brightness of noctilucent clouds is measured on a five-point scale from 1 (weak) to 5 (extremely bright).

Table 8.2. Sample Noctilucent Cloud Observation

Observer: N. M. Bone
Location: Apuldram England 50° 49.8′N 0° 48.3′ W
Date: 6–7 July 2005

UT	Azi	↗	Notes
2158	330–010	10°	NLC types II, III. Brightness 2–3. Bright blue–white patch below Capella.
2208	315–010	8°	NLC types II, III, brightness 3–4. Intense below Capella.
2215	320–015	7°	NLC types IIB, IIIb, brightness 3–4. Strong to west, and below Capella.
2225			Bright at 340 azi, 000 azi, low down.
2230	350	4°	Very intense, to NW of Capella.
	340–025	6°	NLC types IIb, IIIb, V, brightness 3–4. Excellent!
2235			Really bright below Capella; almost gives impression of "shimmering"!
2243	345–005	6°	Very bright (5) to /4° NW of Capella. NLC type II/III, brightness 3–4 elsewhere.
2255	345–005	4°	NLC type IIIb, brightness 3–4. Brightest just west of Capella.
2300	340–345	4°	Fading. NLC type III, brightness 2–3.

Observers should, naturally, take great care to avoid misidentifying cirrus or other tropospheric forms for noctilucent clouds. As a general rule, from the latitudes of northwest Europe and Canada, noctilucent clouds will frequently appear in the region of sky below the bright star Capella: by the time that Capella is clearly visible, cirrus in that region of the sky should certainly be in the Earth's shadow. The distinctive color and structure of noctilucent clouds also aid identification.

It is also usually the case that cirrus clouds become blurry when viewed in binoculars, whereas the sharp wave-forms common in NLC will bear a slightly magnified view well.

Observational reports of noctilucent clouds are welcomed by those organizations that collect auroral sightings. Reports of nights when the observer can say with certainty that there was *no* noctilucent cloud visible from a given location are also of value in assessing long-term patterns of noctilucent cloud behavior.

Photographic Observation

As with the aurora, photography provides the means for rapid, accurate recording of the appearance of noctilucent cloud displays. Results on color slide films can be very attractive. Since the features to be photographed often lie within 15–20° of the horizon, and the illumination of twilight is quite considerable on many nights when noctilucent clouds are present, the photographer is almost obliged to include features of foreground interest, which can also help to illustrate the scale of the display on subsequent viewing. Observers have enjoyed good results with Kodak and Fuji slide films. As a general rule, ISO 400 film works well, with exposures between 2 and 4 seconds, depending on the background sky brightness, at a lens aperture of $f/2.8$.

Since noctilucent clouds will often, from lower latitudes at least, appear largely within a 10° strip above the horizon, the best results will usually be obtained with standard 50 mm focal-length camera lenses, rather than shorter-focus wide angle types. The reduced image scale of the latter means that, although the wide expanse of a display may be recorded in its entirety, little detail is seen on the eventual photograph. Rather more effectively, Danish observers have prepared "mosaics" of several

Figure 8.10.
Noctilucent cloud display on 23–24 June 1995, showing Type III banding. Exposure 3s at $f/2.8$ on Kodak Elite ISO 400 colour slide film. Image: Neil Bone.

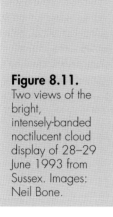

Figure 8.11.
Two views of the bright, intensely-banded noctilucent cloud display of 28–29 June 1993 from Sussex. Images: Neil Bone.

overlapping 50mm exposures. Only when displays are very extensive is a wide angle lens likely to be more useful for recording noctilucent clouds.

Digital cameras are well-suited for recording noctilucent cloud displays. Good pictures can be taken even with quite basic models used in "Night Shot" or similar mode.

Parallactic photography programs, using pairs of photographs taken by observers separated by tens of kilometers, have great potential for further elucidating high atmosphere circulation patterns. Dr. Gadsden suggested that observers at all locations should take photographs of noctilucent cloud features *exactly* on the hour, quarter hours and half hour, on nights when noctilucent cloud is present. The use of fixed mounting brackets at each station allows the aiming direction of the camera to be precisely determined. Such a system can also be used for auroral photography.

Simultaneous occurrences of noctilucent clouds and aurora are infrequent, but not unknown. The detailed effects of auroral activity on noctilucent clouds have yet to

Figure 8.12.
Foreground tropospheric cloud often appears dark in silhouette against bright noctilucent cloud, as in this photograph from 1–2 July 1996. Image: Neil Bone.

Figure 8.13.
Nacreous cloud, photographed on the evening of 16 February 1996 from Wakefield, West Yorkshire. Image: Melvyn Taylor.

be closely studied, thanks mainly to a shortage of observational material. A series of photographs taken on those occasions when noctilucent clouds and aurorae appear together may therefore be of great scientific value. Such observations may show, for example, whether the noctilucent cloud field does indeed become disrupted by possible auroral heating under disturbed geomagnetic conditions.

Also of interest are instances of noctilucent clouds appearing at unusually low latitudes (below 50°N, for example). Several such displays have been photographed by Jay Brausch from latitude 47°N in North Dakota since the summer of 1995, and one notable display in June 2003 was recorded in southern France at latitude 43°N, while reports in the excellent summer of 2006 came from as far south as 44°N in Italy. One sighting, in June 1999, was made from Colorado at latitude 39°N.

Nacreous Clouds

Noctilucent clouds should not be confused with *nacreous clouds* (also known as polar stratospheric clouds), which occur lower in the atmosphere at altitudes between 21 and 30 km. These are, in any case, most frequently seen at high latitudes during the winter months. A strong display was seen over much of the British Isles, as far south as Sussex, on the evening of 16 February 1996. Nacreous clouds, so named for the delicate mother-of-pearl colors which they sometimes exhibit, can be frequently observed from Arctic and Antarctic regions, and have been implicated in processes of ozone depletion.

Airglow

The night sky of the Earth is not perfectly dark. Even in the absence of auroral activity, a weak background of *airglow* suffuses the sky. Airglow results from re-emission of energy from atmospheric particles following daytime excitation by solar radiation.

Light emissions of distinctive character occur during daytime and twilight, too, but the easiest form of airglow for observation is that which occurs at night.

The spectrum of the night-time airglow prominently shows the green (557.7 nm wavelength) line of excited atomic oxygen, but produced by collisions with electrons having energies typically on the order of less than 5 eV, much lower than that of those giving rise to auroral emissions. Triplets of oxygen atoms combine to produce a molecule of oxygen (O_2) and a single excited oxygen atom, which is responsible for the green line emission. Rocket-borne photometers indicate that this emission occurs in a fairly discrete layer 10 km deep around altitudes of 100 km—virtually the same height as the base of the auroral layer. Red oxygen emissions at wavelengths of 630.0 and 636.4 nm occur by more complicated mechanisms at greater heights. As with auroral red emissions, these may only occur where particle densities are sufficiently low that quenching as a result of collisions is not a significant process over the relatively long timescales required for the emission to take place.

Emission lines attributable to sodium become prominent in the twilight airglow, as do the red lines of oxygen. The daytime airglow, although 1000 times more intense than the nightglow, is extremely difficult to observe from the ground thanks to the overall sky brightness. Rocket measurements and the use of extremely narrow-passband filters have allowed dayglow emissions to be isolated from the background for study. Emissions equivalent to those in the nightglow are found at greater intensity in the dayglow, accompanied by more exotic emissions of atomic oxygen and molecular nitrogen which result from collisions between these and free electrons released by solar ultraviolet from other atomic and molecular species.

While weak and diffuse when observed from the ground, airglow can be seen as a fairly strong layer when viewed edge-on in contrast against the blackness of space in photographs from satellites.

The night-time airglow shows variations in brightness, some of which may be associated with geomagnetic activity. Observed increases in red oxygen night-time airglow emissions have been attributed to the arrival, via magnetic field lines, of excited particles from the sunlit conjugate point, for example.

Rocket Launches and Releases from Satellites

Occasional anomalous events are reported, where observers at one locality believe they have seen an auroral display, which cannot be confirmed by reports from elsewhere or by measured geomagnetic activity. Some of these may be attributed to local light pollution, as in the case of what appeared to be a display of auroral rays from south London in August 1988, produced by upwards-pointing searchlights illuminating haze. It may sometimes be less easy to account for other events.

Unusual noctilucent cloud effects may also be associated with rocket launches. A brilliant and multicolored display observed from Finland and Estonia in July 1988 may have been seeded by the exhaust from a Soviet rocket launch. Atmospheric effects produced by rocket launches are frequently reported by Finnish amateur astronomers.

A number of unusual sightings also result from gas releases from orbiting satellites or experimental rockets. Observers in Germany and eastern Europe witnessed a bright diffuse glow produced by solar excitation of exhaust from a US Centaur rocket stage on the night of 3 May 1994. The V-shaped object, which appeared as bright as magnitude −4 (equal in brilliance to the planet Venus) to some observers, initially sparked some "UFO" alerts, before its cause was realized.

Studies of the upper atmosphere, ionosphere, and inner magnetosphere have, since 1955, been carried out using vapor releases from sounding rockets and satellites. Barium is a favored material for this work owing to its low ionization potential. After release, the barium vapor rapidly becomes photoionized by solar ultraviolet. This gives rise to two principal stages in the development of the cloud as observed from the ground. Initially, neutral barium appears as an expanding spherical cloud, emitting by resonance scattering at 553.3 nm in the green. Photoionization leads to violet emissions at 493.4 and 455.4 nm wavelengths. The cloud then becomes elongated as the barium ions spread along (but not *across*) geomagnetic field lines. This striated form can appear similar to auroral rays.

The development of barium clouds following release can be monitored using low-light television systems and photography. The violet emissions can be recorded on color film, but may more typically appear dull grey to the eye.

Barium clouds may be quite readily seen with the naked eye, provided the Sun is at least 8° below the observer's horizon while remaining above the horizon at the atmospheric level where the release has occurred. Most barium release experiments have been carried out in the ionosphere between 140 and 400 km altitude. Barium cloud tracing of the ionosphere overlying existing auroral arcs during displays allows investigation of the region in which auroral electrons undergo their final acceleration. While barium cloud structures can resemble auroral arcs, the particles within the former are much less energetic.

Barium releases have occasionally been made over more populous areas where aurorae might also be visible, but advance warning has usually been given in the scientific literature, enabling amateur observers to add their support in monitoring cloud development. A number of barium cloud releases have been made from rockets launched at the Poker Flat range in Alaska. Releases from sounding rockets launched from the Hebridean island of South Uist in the early summer of 1973, for example, were followed by amateur observers in central Scotland.

In the late 1960s and early 1970s, experiments were carried out using rocket-borne electron accelerators to trace out geomagnetic field lines. Pulses of electrons, accelerated to energies of 10 keV (similar to those found in aurorae), produced sub-visual excitation of the atmosphere, detectable using low-light television cameras. Electron beams have been successfully fired from one conjugate point to the other with minimal energy loss. Shaped-charge barium releases have also been used to trace geomagnetic field lines, again with the aid of low-light cameras.

Experimental barium releases have also taken place outside the magnetosphere, to investigate plasma motions in the solar wind. These have included the AMPTE and CRRES experiments (Chapter 7); some of the latter's gas releases were observed and photographed by amateur astronomers in the United States during January 1991.

In the "cold-war" climate of the late 1950s and early 1960s, several nuclear weapon tests were carried out at altitude in the atmosphere by both the Soviet Union and United States. The effects of these test explosions were spectacular, and long-lived. Temporary belts of enhanced particle density and radiation were produced in the plasmasphere region, with deleterious effects for many orbiting satellites, which suffered damage to their solar cells. Contamination of the magnetosphere by the particles released during these explosions persisted for many years, making measurements of the natural state of the near-Earth environment difficult.

Associated with the explosions of such as Project Starfish, launched from Johnston Island in the Pacific Ocean on 9 July 1962, were auroral-type displays, visible at either conjugate point. The Starfish explosion provided a brief, spectacular display for observers in New Zealand.

The Zodiacal Light

A phenomenon occurring far beyond Earth's high atmosphere, the diffuse glow of the *zodiacal light* is produced mainly by reflection of sunlight from myriads of small particles in the inner Solar System. These particles, whose size is typically from 0.1 to 10 microns, are believed to originate principally from comets and asteroids. Estimates suggest that the inner Solar System dust cloud contains about one cometary mass of these particles. Solar radiation pressure interactions (the Poynting–Robertson effect) result in a steady depletion of material from the dust cloud, such that without continued replenishment from new comets and asteroid fragment collisions it would disappear within a few thousands of years.

Observations obtained using photopolarimeters aboard the Pioneer spacecraft in the early 1970s indicated that the solar system dust cloud is shepherded by Jupiter's gravitational field such that the zodiacal light and *gegenschein* (counterglow) are absent outside the giant planet's orbit.

The zodiacal light and associated skyglows are increasingly difficult to observe as light pollution spreads with urban growth. Few present-day astronomers have seen the zodiacal light, while the *gegenschein* is even more elusive. These glows were, however, known to astronomers at least as far back as the seventeenth century when Cassini carried out systematic observations of the zodiacal light, and noted that it varied in brightness from time to time.

The zodiacal light's brightness variations may be correlated with the solar cycle. It has been suggested that the zodiacal light is brighter at sunspot minimum than at sunspot maximum. A probable explanation for this is the presence of energetic coronal hole particle streams permeating the solar system in the years around sunspot minimum, producing excitation of the tenuous interplanetary medium. Brightening of the zodiacal light on short timescales of a few days has also been reported by some observers, and may be connected to solar flare events.

Typically, the zodiacal light appears as a broad cone of diffuse light, little brighter than the Milky Way, and is best seen from temperate latitudes either in the evening sky about 90 minutes after sunset around the spring equinox, or in the morning sky around the autumnal equinox. At other times, it cuts a shallow angle relative to the horizon and is lost among the haze. The zodiacal light is seen to best advantage from the tropics, where the angle of the ecliptic to the horizon is steep. The faint *gegenschein* is only visible in the very darkest of skies, appearing as a diffuse oval of reflected sunlight some 10° or 20° in diameter in the midnight sky directly opposite the Sun.

The Auroral Sound Controversy

As a phenomenon of the high atmosphere, the aurora should, sensibly, be so far distant from the observer that no sound can be heard. Several witnesses, however, are adamant that they have heard sound during times of intense auroral activity, occurring simultaneously with visual outbursts—often as arcs or bands pass overhead during the breakup phase of substorm displays, for example (Chapter 4). The interval during which sounds are heard by such witnesses, who have included professional scientists engaged in auroral research at high latitude sites, is typically short, perhaps of the order of 10 minutes.

Descriptions of the sounds heard include faint whistling, rustling, swishing, and crackling or a soft hissing. Similar sounds are sometimes reported in association with meteors. As with auroral sounds, reports of simultaneous sounds from meteors are, in general, greeted with considerable skepticism by the scientific community.

There are two obvious problems with simultaneous auroral (or meteor) sound. First, the aurora occurs in near vacuum, mostly in the tenuous upper atmosphere above 100 km altitude, such that there is very little medium through which sound waves can be propagated downwards. Perhaps more critically, the speed of sound in the atmosphere is such that any noises produced at auroral altitudes should take a matter of minutes to reach ground level, making it hard to reconcile the occurrence of auroral sounds simultaneous with visual activity.

A number of explanations have been offered to account for the possibility of auroral sound. One, physiologically based, suggestion is that the sounds are simply a consequence of the ears' adjustment to cold outdoor conditions. This does not, however, account for the occasions on which witnesses report *no* sounds under conditions of high activity. Alternatively, the effect may merely be psychological, with observers sub-consciously *expecting* to hear sounds associated with vigorous activity.

Among the physical mechanisms proposed, propagation of VLF radio waves, and their more-or-less instantaneous re-transmission from suitable nearby "receivers," has enjoyed some support, both for auroral and meteor sounds. A more detailed investigation of VLF propagation from very bright meteors (fireballs) by S. M. Silverman and T. F. Yuan and colleagues, however, has cast doubt on this electrophonic mechanism. An alternative is *coronal discharge* from pointed objects (such as the tips of coniferous trees) in the vicinity of the observer; St. Elmo's Fire is perhaps the most familiar example of this sort of phenomenon, usually associated with thunderstorms. Coronal discharge requires the generation of strong electrical fields, which may be correlated with auroral activity. This mechanism is dependent on the dryness of the air, perhaps offering an explanation for the reporting of simultaneous sound associated with some auroral displays, but not others.

While hundreds of reports of anomalous auroral sound have been made since the eighteenth century, it is perhaps significant that attempts to record these sounds using modern equipment have produced only ambiguous results. The occurrence or otherwise of auroral sound is likely to remain contentious for some time to come.

Appendix: Observational Organizations

Aurora and noctilucent cloud observations are collected by the British Astronomical Association via its Aurora Section, which also issues standard guidelines for visual observations. The Section can be contacted through:

The British Astronomical Association,
Burlington House,
Piccadilly,
London W1V 9AG
http://www.britastro.org

The Finnish Ursa Astronomical Association has an active aurora observing group:

Laivanvarustajakata 9 C 54,
01400 Helsinki,
Finland

In Germany, the AKM group has been active in collecting aurora and noctilucent cloud sightings:

Jurgen Rendtel,
AKM,
PF 600118,
D-14401 Potsdam,
Germany

In the southern hemisphere, observations can be sent to the Royal Astronomical Society of New Zealand:

Royal Astronomical Society of New Zealand,
PO Box 3181,
Wellington,
New Zealand

North American and Canadian noctilucent cloud observations are collected by NLC CAN-AM:

Mark Zalcik,
9022–132 A Ave,
Edmonton, Alberta,
T5E 1B3
Canada

Radio observers of auroral and other effects may find contact with the national organizations in that field profitable:

Radio Society of Great Britain,
Lambda House,
Cranborne Road,
Potter's Bar,
Herts., EN6 3JE

The RSGB broadcasts forecasts of likely radio auroral conditions for the week ahead on GB2RS each Sunday.

American Radio Relay League,
Newington,
CT 06111, USA

Glossary

aa index: Measure of the overall level of geomagnetic activity from equivalent (antipodal) locations in opposite hemispheres, based on 12-hour averages.

Airglow: Low-level diffuse light emission across the whole sky, resulting from excitation of atmospheric particles by solar radiation during daytime.

Ap index: Measure of geomagnetic activity based on *magnetometer* records from 12 observatories worldwide. Similar to the *Kp index*, but the Ap index is linear (rather than logarithmic) and based on 24-hour averages.

ap index: A 3-hourly average of geomagnetic activity on a linear scale similar to the *Ap index*. The ap index can be broadly related to *Kp index* values.

Arc: "Rainbow" or arch-shaped form of *discrete aurora*, usually aligned east–west along the *auroral oval*. From lower latitudes, arcs are seen to have their highest point around the direction toward the magnetic pole. Arcs may be homogeneous or rayed. The lower edges of arcs are usually more sharply defined than the upper.

Auroral breakup: Main phase of vigorous auroral activity, dominated by moving *arcs*, *bands*, and *rayed* forms, during a *substorm*.

Auroral Kilometric Radiation (AKR): Radio emission from the aurora at 100 kHz to several hundred kHz frequency, detectable from space.

Auroral ovals: Rings of auroral activity around either geomagnetic pole. Aurora is found more or less permanently around these rings, which are displaced toward Earth's night-side, having their greatest equatorwards extent around the midnight point. Under quiet conditions, the auroral ovals have a diameter of 4000–5000 km. *Geomagnetic storms* cause the ovals to expand markedly, particularly on the night-side.

Auroral potential structure: Thin sheets of positive and negative charge, aligned to magnetic field lines and lying 10,000–20,000 km above auroral heights in the atmosphere. Electrons undergo acceleration along these thin sheets, gaining sufficient energy to produce auroral excitation on impact with oxygen and nitrogen in the upper atmosphere.

Auroral zone: Circular region around either geomagnetic pole traced out by the maximum equatorwards extent of the *auroral ovals*, and within which aurorae are most likely to be seen. The northern hemisphere auroral zone crosses the North Cape of Norway, Iceland, northern Canada, Alaska and Siberia.

Band: Twisted, ribbon-like form of *discrete aurora*, often resulting from folding of an *arc*, and usually aligned east–west along the *auroral oval*. Bands may be homogeneous or rayed. Formation of, and movement within, bands often signals the onset of increased auroral activity. The lower edges of bands are usually more sharply defined than the upper.

Bartels diagram: A means of presenting auroral activity data in 27-day strips, each corresponding to a single rotation of the Sun on its axis as seen from Earth. Recurrent aurorae due to persistent solar active features (such as *coronal holes*) become evident by their alignment in adjacent strips.

Bow shock: Wave front produced "upstream" of a body in the *solar wind*, analogous to that produced by a ship ploughing through water. Solar wind plasma is decelerated and deflected by passage through the bow shock, and then flows around the *magnetopause*. Bow shocks are seen upwind of the major planets, and ahead of comets.

Chromosphere: Inner region of the Sun's atmosphere, visible at total solar eclipses as a red ring. The chromosphere can be routinely studied at the wavelength of hydrogen-alpha light. The chromosphere lies above the *photosphere*, reaching to about 10,000 km above the solar surface. Overlying the chromosphere is the *corona*. Chromospheric temperatures are typically of the order of 10,000K.

Cleft: Region in the high-latitude dayside of the *magnetosphere* within which magnetic field lines are closely bundled together. The cleft surrounds the narrower *cusp*.

Conjugate points: Locations of equivalent magnetic latitude and longitude in either hemisphere.

Corona: (a) The bright inner atmosphere of the Sun, visible during total solar eclipses. Plasma in the corona is at extremely high temperatures (1,000,000 K). (b) Form assumed by the aurora when features pass overhead from the observer's location on the ground. As a result of perspective, rays and other features appear to fan out from a central point.

Coronal hole: Region of open magnetic field in the Sun's inner atmosphere, from which a high-speed particle stream emerges into the *solar wind*. Coronal holes and their associated streams may last for several months, giving rise to recurrent magnetic disturbances at 27-day intervals.

Coronal Mass Ejection (CME): Expanding "bubble" structure seen moving outwards through the Sun's corona at times of high solar activity, believed to be the result of material being thrown out from the inner solar atmosphere following magnetic reconnection. CMEs are associated with solar flares and the disappearance of filaments.

Coronal transient: Equivalent to a *coronal mass ejection*.

Cosmic rays: Not rays at all, but subatomic particles—ions—that have undergone acceleration in violent cosmological events such as supernovae.

Crotchet: "Saw-tooth" feature produced on a *magnetogram* as a consequence of dayside Sudden Ionospheric Disturbance events. The brief increase in *D-region* ionization induces an abrupt shift in the ground-level magnetic field, followed by fairly rapid recovery over an hour or so.

Cusp: Narrow high-latitude region on the dayside of the *magnetosphere*, where terrestrial magnetic field lines are closely bundled together and dip sharply toward Earth's surface. *Solar wind* plasma is able to penetrate the magnetosphere via the cusps in either hemisphere.

D-region: Lowest part of the *ionosphere*, between altitudes of 65–80 km.

Diffuse aurora: Structureless auroral light which may accompany *discrete* forms.

Discrete aurora: Clearly identifiable auroral structure, such as *arcs* or *bands*, which may be *rayed* or *homogeneous*.

E-cross-B drift: Motion of a charged particle perpendicular to both electrical current and magnetic field.

E-layer: Region of the *ionosphere* around 110 km altitude.

Earth-radii: A convenient unit of measurement for describing distances within Earth's *magnetosphere*. Earth's equatorial radius is 6370 km.

Electrojets: Electron currents flowing eastwards in the evening sector, and westwards in the morning sector, of the *auroral ovals* at altitudes around 100 km. Where these meet, around the midnight point, a region of turbulence (the *Harang discontinuity*) is produced.

F-layer: The highest part of the *ionosphere*, showing a split into two parts. The F_1 layer at 160 km altitude is a permanent feature, while the F_2 layer around 300 km altitude shows a diurnal variation, disappearing at night.

Filament: Equivalent to a prominence seen dark by contrast in hydrogen-alpha light while in transit. Filaments may disappear following *solar flare* activity elsewhere on the Sun's leading to mass ejection into the *solar wind*. Disappearing filaments may give rise to enhanced geomagnetic activity.

Flaming: Very rapid brightness variation in aurora, wherein waves of brightening sweep upwards from the horizon to the top of the display at a rate of several per second. Flaming may be the prelude to formation of a corona in a very major display, but is also common in the declining phase of a short outburst.

"Flash aurora": As yet poorly understood phenomenon, reported by several experienced observers, where auroral *arcs*, *bands*, or other structures appear in the sky for only a matter of a few seconds before disappearing again.

Forbidden transitions: Electron transitions to higher energy levels surrounding an atomic nucleus which are not normally permitted by quantum mechanical rules. Excitation during auroral conditions leads to forbidden transitions in oxygen and nitrogen.

Forbush decrease: Interlude of diminished Galactic cosmic ray flux at Earth, resulting from passage of a *coronal mass ejection*.

Geocorona: Cloud of neutral hydrogen from the upper atmosphere, surrounding the Earth to a distance of a few *Earth-radii*.

Geomagnetic latitude: Latitude of a given location with respect to the nearer of the two geomagnetic poles. As a result of the offset of Earth's magnetic and geographical axes, geomagnetic latitude is not equivalent to geographical latitude.

Geomagnetic storm: Period during which a major disturbance of the terrestrial magnetic field is brought about following a *solar flare* or *coronal mass ejection*. Magnetic energy and plasma from the *solar wind* disturbs the *magnetotail* plasma population, leading to injection of accelerated particles into the upper atmosphere, and an intensification of auroral activity. A major geomagnetic storm may last several days. Such events, which bring the aurora to lower latitudes during expansion of the *auroral ovals*, are most common at times of high sunspot activity.

Ground Induced Current (GIC): Electrical current at Earth's surface, resulting from electron motions in the ionosphere during a geomagnetic disturbance.

Ground Level Enhancement (GLE): Interlude of increased solar cosmic ray flux at Earth's surface following a *solar flare* or *coronal mass ejection*.

Ground state: Minimum-energy configuration of electrons surrounding an atomic nucleus.

Harang discontinuity: Turbulent region around the midnight point on the *auroral oval* where the eastwards and westwards *electrojets* meet. Poor short-wave radio communications prevail when a given operating station is closest to the auroral oval at midnight.

Heliopause: Boundary between the *heliosphere* and interstellar space.

Heliosphere: Volume of space within which the solar magnetic field and plasma are dominant, extending perhaps 110–120 AU from the Sun, and thus encompassing the realm of the known planets.

Homogeneous: Description given to either *discrete* or *diffuse aurora* in which no internal structure is evident.

Interplanetary Magnetic Field (IMF): Magnetic field, whose characteristics are defined by that of features at the Sun's surface, carried by the *solar wind*. The IMF has a typical strength of 5 nT at the orbit of the Earth. When the IMF polarity is turned southwards with respect to the ecliptic plane, reconnection between the solar wind and Earth's magnetosphere is most efficient, leading to enhanced geomagnetic and auroral activity.

Ionosphere: That part of Earth's upper atmosphere between about 60 and 300 km altitude in which layers of ionization (produced by solar radiation) are found.

Isochasm: Line connecting geographical locations which enjoy the same average annual frequency of aurorae.

Kp index: Measure of the overall level of geomagnetic activity, based on *magnetometer* readings from 12 observatories worldwide. Averages are given for 3-hour intervals. The Kp index is semi-logarithmic. Kp index <5 corresponds to quiet conditions, Kp >5 indicates storm conditions.

Lorentz force: Force operating on charged particles in a magnetic field, whereby these are deflected at right-angles to the field, and at right-angles to their previous direction of motion. This leads to spiral trajectories for protons and electrons traveling along magnetic field lines.

Magnetic sectors: Semipermanent large-scale regions in the solar wind, separated by the neutral sheet, within which the Interplanetary Magnetic Field points either towards or away from the Sun.

Magnetogram: A trace of variations in the local magnetic field as a function of time, as recorded by a *magnetometer*.

Magnetometer: A device for measuring variations in the strength (H in the horizontal direction, Z in the vertical) or the angular deviation (D) of the local magnetic field. Enhanced geomagnetic activity can be detected by virtue of marked variations in H, Z, or D resulting from currents in the ionosphere.

Magnetopause: Outer boundary of a planet's *magnetosphere*.

Magnetosheath: Plasma-containing region surrounding Earth's *magnetosphere* between the *magnetopause* and the *bow-shock*.

Magnetosphere: Comet-shaped volume of space in which a planet's magnetic field is dominant over that of the surrounding *solar wind*.

Magnetotail: The extended "downwind" section of a planet's *magnetosphere*. Earth's magnetotail reaches far beyond the Moon's orbit on the planet's night-side.

Maunder Minimum: Period of apparently diminished sunspot (and auroral) activity between about 1645 and 1715.

Mirror point: Point in the trajectory of a charged particle moving along a magnetic field at which its motion is exactly perpendicular to the field line. At this point, the charged particle is deflected back along the field line in the opposite direction. Particles in the *Van Allen belts*, for example, are deflected back and forth between mirror points in opposite hemispheres, thereby being "trapped."

Neutral sheet: A feature of the extended solar atmosphere lying roughly in the Sun's equatorial plane, separating hemispheres of north and south magnetic polarity. Particularly at times of high solar activity, this may become "pleated," rather than lying in a flat plane.

Noctilucent clouds: Tenuous clouds, probably comprising water ice condensed onto small particles of meteoric origin, forming close to the mesopause around 82 km altitude. Noctilucent clouds are a summer phenomenon, seen under twilight conditions when the Sun is between 6° and 16° below the observer's horizon, usually from latitudes higher than 50° (reports of sightings to as low as 45° have been made in recent years).

Photosphere: The bright visible surface of the Sun. Photospheric temperatures are typically of the order of 600 K, being reduced to around 4000 K in sunspots, which appear dark in contrast.

Plasma: State of matter in which a gas is completely ionized and highly electrically conductive.

Plasma mantle: Population of plasma flowing along the outside of the lobes in Earth's *magnetotail* on the boundary of the *magnetosheath*.

Plasma sheet: Region of Earth's *magnetosphere* lying between the opposed north and south magnetic polarity lobes of the *magnetotail*. Plasma near the plasma sheet may be injected into the upper atmosphere, causing aurora, at times of high geomagnetic activity.

Plasmasphere: Region of Earth's *magnetosphere*, containing low-energy plasma lying under the outer *Van Allen belt* to a distance of 4 *Earth-radii*. This plasma population is derived from material lost from the ionosphere.

Plasmoids: Pockets of plasma from the *magnetosphere*, enclosed by magnetic "cage" structures, and ejected down the *magnetotail* into the *solar wind* at times of enhanced geomagnetic activity.

Polar Cap Absorption (PCA): Auroral event occurring polewards of the auroral oval, resulting from the arrival of high-energy protons in the *solar wind*. Associated with PCA is disruption of short-wave radio communication at high latitudes, and weak (often sub-visual) *polar glow aurora*.

Polar glow aurora: Weak, diffuse auroral activity present within the *auroral ovals* during *Polar Cap Absorption* events.

Prominence: Loop of gas from the *chromosphere* reaching perhaps as much as 50,000 km above the Sun's surface, and into the *corona*: prominences are sometimes visible during total solar eclipses as red "flames" surrounding the dark body of the Moon. Prominences are supported by the solar magnetic field. In the light of hydrogen-alpha, prominences may also be seen as dark *filaments* against the bright background.

Pulsation: Slow brightness variation within aurora, taking place over timescales from a few seconds to several minutes.

Radio aurora: Enhanced short-wave communication conditions, resulting from increased ionization in the *E-layer* during a geomagnetic disturbance. Radio auroral events do not always coincide with visual displays, and vice versa.

Rayed: Description given to forms of *discrete aurora* in which vertical structure is present. Rays may extend for some considerable distance above the base of an *arc* or *band*. Isolated ray bundles, bearing resemblance to searchlight beams, may also be seen on occasion.

Resonance scattering: Absorption, by molecule or atom, of sunlight at a specific wavelength and its subsequent re-emission at the same wavelength. Found, for example, in sunlit aurora, where ionized molecular nitrogen (N_2^+) absorbs and re-emits solar radiation at 391.4 nm, resulting in enhanced blue–purple emission.

Sector boundary crossing: Interval during which Earth (or another body) passes through the pleated neutral sheet in the solar wind from a region of one magnetic polarity to a region of opposed polarity. Sector boundary crossings may add energy to the auroral dynamo, giving a brief enhancement in activity.

Solar constant: Total energy output of the Sun, as measured from satellites above the atmosphere (and found to be anything *but* constant!).

Solar flare: Violent outburst, involving the release of magnetic energy, occurring in the inner solar atmosphere above sunspot groups. Flares are accompanied by ejection of material into the *solar wind*, whose arrival at Earth 24–36 hours later may trigger a *geomagnetic storm*. Flares may be observed at the wavelengths of hydrogen-alpha light, and also produce bursts of radio noise.

Solar radiation storm: Interval of elevated proton ('solar cosmic ray') flux in near-Earth space, following a major solar flare. These are described in terms of intensity from S1 to S4, the later being severe. Equipment aboard artificial satellites may be damaged by particle 'hits' during solar radiation storms.

Solar wind: continuous outflow of plasma from the Sun. Under quiet conditions, the solar wind flows past Earth at 400 km/s. Following *coronal mass ejections* or *solar flares*, local solar wind velocities of 1000–1400 km/s may be found. *Coronal holes* introduce streams of solar wind flowing at 600–800 km/s into the solar wind.

South Atlantic Anomaly: Region above the Atlantic Ocean off the coast of Brazil where the inner *Van Allen belt* reaches a minimum altitude of 250 km above Earth's surface. Artificial satellites passing through this region experience increased exposure to energetic particles.

Sporadic E: Thin sheets of enhanced ionization in the *E-layer* of the *ionosphere* frequently found during daytime in the summer. Sporadic E causes disruption of short-wave radio communication over areas of 1000–2000 km. The mechanism by which Sporadic E arises is as yet poorly understood.

Substorm: Disturbance of the *auroral ovals* resulting from increased energy input from the *magnetotail*. During substorms, the oval brightens, then expands, on the night-side. Substorms may occur one to three times daily under normal conditions, and more frequently at times of high solar activity.

Sudden Ionospheric Disturbance (SID): Rapid increase in the level of ionospheric *D region* ionization resulting from X-ray emissions associated with *solar flares*. The increased ionization causes short-wave radio fade-outs, lasting of the order of a few minutes to an hour, on Earth's day-side.

Sudden Storm Commencement (SSC): Abrupt onset of disturbed geomagnetic conditions, marking the arrival of an interplanetary shock wave associated with a *coronal mass ejection*. During SSC, the horizontal field strength measured by a *magnetometer* may intensify markedly as a result of compression of the *magnetosphere*. The local magnetic field my also show marked angular deviation. SSC events are often, but not always, followed by *geomagnetic storms*.

Theta aurora: Configuration of the *auroral oval* at times of low geomagnetic activity when a transpolar arc of *discrete aurora* links the day- and night-sides of the oval along the noon-midnight line. So called for it resemblance to the Greek letter.

Van Allen belts: Regions within Earth's *magnetosphere* in which energetic particles are trapped. The inner belt has an equatorial distance of 1.5 *Earth-radii* and contains protons and electrons of both terrestrial and *solar wind* origin. The outer belt is populated mainly by electrons from the solar wind and has an equatorial distance of 4.5 Earth-radii.

Zodiacal light: Diffuse glow, seen as a cone of faint light extending along the ecliptic in the late evening or pre-dawn, resulting from reflection of sunlight from small particles in the plane of the Solar System. The zodiacal light's brightness may be enhanced from time to time by energetic particle streams in the *solar wind*.

Bibliography

Akasofu, S.-I., and Kamide, Y. (Eds.) *The Solar Wind and the Earth.* D. Reidel (1987).

Beatty, J. K., Petersen, C. C., and Chaikin, A., *The New Solar System* 4th Edition. Sky Publishing (1998).

Bone, N., *Observer's Handbook Meteors.* Philip's (1993).

Brekke, A., and Eggeleand, A., *The Northern Lights, Their Heritage and Science.* D. Reidel (1994).

Carlowicz, M. J., and Lopez, R. E., *Storms from the Sun.* Jospeh Henry Press (2002).

Davis, N., *The Aurora Watcher's Handbook.* University of Alaska Press (1992).

Eather, R. A., *Majestic Lights: the Aurora in Science, History and the Arts.* American Geophysical Union (1980).

Freeman, J. W., *Storms in Space.* Cambridge University Press (2001).

Gadsden, M., and Schroeder, W., *Noctilucent Clouds.* Springer Verlag (1989).

Greenler, R., *Rainbows, Halos and Glories.* Cambridge University Press (1989).

International Union of Geodesy and Geophysics, *International Auroral Atlas.* Edinburgh University Press (1963).

McDonald, L., *How to Observe the Sun Safely.* Springer (2003).

Meng, C.-I., Rycroft, M. J., and Frank, L. A., (Eds.) *Auroral Physics.* Cambridge University Press (1991).

Minnaert, M., *The Nature of Light and Colour in the Open Air.* Dover (1954).

Newton, C., *Radio Auroras.* Radio Society of Great Britain (1991).

Odenwald, S., *The 23rd Cycle: Learning to Live with a Stormy Star.* Columbia University Press (2001).

Philips, K. J. H., *Guide to the Sun.* Cambridge University Press (1992).

Wentzel, D.T., *The Restless Sun.* Smithsonian University Press (1989).

Numerous sources of up-to-date observational material exist, of which the best are probably the websites and newsletters of the respective observational organizations. Reports of activity from within the previous six weeks can be found in *The Astronomer,* published monthly (http://www.theastronomer.org.uk). *Sky and Telescope, Astronomy Now,* and *Astronomy* all carry reports from time to time, though these can sometimes be lacking in useful observational details such as dates and places! These magazines also carry news of the latest spacecraft missions to investigate solar–terrestrial relations.

Index

Other Titles in this Series

(*continued from page ii*)